SpringerBriefs in Applied Sciences and Technology

Computational Intelligence

Series Editor

Janusz Kacprzyk

For further volumes:
http://www.springer.com/series/10618

Shilpa Metkar · Sanjay Talbar

Motion Estimation Techniques for Digital Video Coding

 Springer

Shilpa Metkar
Electronics and Telecommunications
College of Engineering
Pune, Maharashtra
India

Sanjay Talbar
Electronics and Telecommunications
Shri GGS Institute of Engineering
and Technology
Nanded, Maharashtra
India

ISSN 2191-530X   ISSN 2191-5318   (electronic)
ISBN 978-81-322-1096-2   ISBN 978-81-322-1097-9   (eBook)
DOI 10.1007/978-81-322-1097-9
Springer New Delhi Heidelberg New York Dordrecht London

Library of Congress Control Number: 2013932556

Printed on acid-free paper

Springer is part of Springer Science+Business Media (www.springer.com)

To my
Passion and Stimulus
The one who is behind my crafted
achievements

Shilpa Metkar

To my
Parents, my wife—Rohini and son
Saurabh and Shubham

Sanjay Talbar

Preface

Motion Estimation (ME) and compensation techniques, which can eliminate temporal redundancy between adjacent frames effectively, have been widely applied to popular video compression coding standards such as MPEG-2 and MPEG-4.

The communication area has been widely benefited by the developments in signal processing, which enabled variety of services in video sector.

With the increasing popularity of technologies such as Internet streaming video and video conferencing, video compression has became an essential component of broadcast and entertainment media. Video signal processing means: to compress the video frame; to encode the compressed frame; and then transmit the same. At the receiving end the video is displayed by decoding of an arriving signal. The aim of the video signal processing should be to compress maximum data in a minimum possible bandwidth. There are number of techniques to accomplish this process with their respective advantages and disadvantages.

This book incorporates comprehensive account of techniques suggested by different people for achievement of a more faithful transmission and reproduction of videos using limited bandwidth. The author has attempted to present different techniques starting with most initial to most advanced and effective, so that the student will be able to learn them with ease.

Our goal for this book is to provide an extensive development of a methodology to estimate the motion field between two frames pertaining to video coding applications.

Block matching techniques are generally used for motion estimation in video coding. In this context, the best solution from the quality point of view is represented by a full search algorithm that considers every possible detail while requiring however an enormous computational complexity. Different suboptimal solutions have been proposed in the literature. This book proposes an exhaustive study of the motion estimation process in the framework of a general video coder.

This work presents a novel method named as Modified Orthogonal Search Algorithm (MOSA) for the block-based motion estimation. We introduced the center-biased search point pattern for the estimation of small motions and a half way stop technique to reduce the computational complexity.

Algorithmic analysis shows that motion estimation is the most complex module in the video encoder. This is mainly due to the involvement of great number of calculations for motion estimation. Having this in mind, this book presents an innovative algorithm, for a further complexity reduction of the motion estimation (ME) module of video coder, by employing motion detection prior to motion estimation. Simulation results of the proposed technique reported a very good improvement in reducing the computations. An early detection of blocks due to zero motion vector, leads to cut redundant computation significantly, which speeds up the coding of video sequences.

The overall structure of this book takes the form of five chapters.

Chapter 1 provides a concise introduction to the video compression techniques. It explains the main motivation for the research described in this book.

A theoretical and practical framework for the existing block matching algorithms is described in Chap. 2. The chapter also highlights modification suggested in the existing algorithm. This chapter also covers the video sequences used in this work for the analysis and simulation.

Chapter 3 presents a novel method named as Modified Orthogonal Search Algorithm (MOSA) for the block-based motion estimation. The advantage of the center-biased search point pattern for the estimation of small motions and a halfway-stop technique to reduce the computational complexity is elaborated. Detail analysis of the result is discussed stating its advantages over existing algorithm.

A novel reduced complexity motion estimation technique is introduced in Chap. 4.

Chapter 5 finally concludes this work by giving a brief summary and critique of the findings as well as identifying areas for future research.

Acknowledgments

I feel very proud and honored today, for fulfilling the dreams of my parents. I take this opportunity to express my gratitude towards them, for making me predisposed of all these achievements. Dr. S. N. Talbar my teacher, guide and philosopher. I offer complete credit of all my academic achievements to him for carving my personality.

I am thankful to all my teachers of S.G.G.S.I.E. and T. Nanded for their guidance and counseling during my academics. I thankful to all my faculty friends from College of Engineering, Pune for their encouragement and support. I had a deep feeling of gratefulness towards Dr. [Mrs.] M. A. Joshi for her love and affection. I am thankful to Hon. Director Dr. Sahastrabudhe A. D. for inspiring me to take this project.

It would be injustice with all my students, if I did not express my warmth to them. I am highly obliged to all family members and friends for their continuous support. Lastly thanks are due to my passion, which always stimulate me to be instrumental.

Shilpa Metkar

I would like to express my heartfelt thanks to the faculty members of the Electronics and Telecommunication Engineering Department, Shri Guru Gobind Singhji Institute of Engineering and Technology, Nanded for their appreciation.

I am grateful to Dr. L. M. Waghmare, Hon. Director Shri Guru Gobind Singhji Institute of Engineering and Technology, Nanded for his guidance and encouragement.

I articulate my indebted towards my students.

I express my gratitude to my family for their support and enthusiastic cooperation. Lastly, I thank God almighty for providing me strength.

Sanjay Talbar

Contents

Chapter 1
Introduction

Digital video compression techniques have played an important role in the world of telecommunications and multimedia systems where bandwidth is still a valuable commodity. Video compression has two important benefits. First, it makes it possible to use digital video in transmission and storage environments that would not support uncompressed ('raw') video. For example, current Internet throughput rates are insufficient to handle uncompressed video in real time (even at low frame rates and/or small frame size). A Digital Versatile Disk (DVD) can only store a few seconds of raw video at television-quality resolution and frame rate and so DVD-Video storage would not be practical without video and audio compression. Second, video compression enables more efficient use of transmission and storage resources. If a high bit rate transmission channel is available, then it is a more attractive proposition to send high-resolution compressed video or multiple compressed video channels than to send a single, low-resolution, uncompressed stream. Even with constant advances in storage and transmission capacity, compression is likely to be an essential component of multimedia services for many years to come.

1.1 Video Signal Formats

Exchange of digital video between different industries, applications, networks, and hardware platforms requires standard digital video formats. There are many digital video formats accepted for different applications and they are usually related to the CCIR 601 sampling standard [1]. CCIR-601 was defined mainly for broadcast-quality applications. For storage applications, a lower resolution format called the Source Input Format (SIF) was defined. A lower resolution version of SIF is the quarter-SIF (QSIF) format.

The recommendation H.261 from CCITT defines another source format, named common intermediate format or CIF and Quarter-CIF (QCIF) [2]. A set of popular frame resolutions is based on the CIF, in which each frame has a resolution of

S. Metkar and S. Talbar, *Motion Estimation Techniques for Digital Video Coding*, SpringerBriefs in Computational Intelligence, DOI: 10.1007/978-81-322-1097-9_1, © The Author(s) 2013

352×288 pixels. The resolutions of these formats are listed in Table 1.1. Other video formats including CCIR 601, SIF, and HDTV are also summarized in Table 1.1.

1.2 Video Standards

Since there are endless ways to compress and encode data, common standards are needed that rigidly define how the video is coded in the transmission channel. There are mainly two standard series in common use, both having several versions. The International Telecommunications Union (ITU) started developing Recommendation H.261 in 1984, and the effort was finished in 1990 when it was approved. The standard is aimed for video conferencing and video phone services over the integrated service digital network (ISDN) with bit rate as a multiple of 64 kbits/s. MPEG-1 is a video compression standard developed in joint operation by International Standards Organization (ISO) and International Electro-Technical Commission (IEC). The system development was started in 1988 and finished in 1990, and it was accepted as standard in 1992. MPEG-1 can be used at higher bit rates than H.261, at about 1.5 Mbits/s, which is suitable for storing the compressed video stream on compact disks or for using with interactive multimedia systems [3]. The standard also covers audio associated with a video.

In 1996 a revised version of the standard, Recommendation H.263, was finalized which adopts some new techniques for compression, such as half pixel and optionally smaller block size for motion compensation. As a result it has better video quality than H.261. Recommendation H.261 divides each frame into 16×16 picture element (pixel) blocks for backward motion compensation, and H.263 can also take advantage of 8×8 pixel blocks. A new ITU standard in development is called H.26L, and it allows motion compensation with greater variation in block sizes.

For motion estimation, MPEG-1 uses the same block size as H.261, 16×16 pixels, but in addition to backward compensation, MPEG can also apply bidirectional motion compensation. A revised standard, MPEG-2, was approved in 1994. Its target is at higher bit rates than MPEG-1, from 2 to 30 Mbits/s,

Table 1.1 Parameters of various video formats	Format	Luminance resolution (*horizontal* × *vertical*)
	CCIR-601	720×486
	SIF	352×240
	QSIF	176×120
	CIF	352×288
	QCIF	176×144
	HDTV	1920×1080

where applications may be digital television or video services through a fast computer network. The latest ISO/IEC video coding standard is MPEG-4, which was approved in the beginning of 1999. It is targeted at very low bit rates (832 kbits/s) suitable for, e.g., mobile video phones. MPEG-4 can be also used with higher bit rates, up to 4 Mbits/s.

More recently, MPEG-7, formally named 'Multimedia Content Description Interface', concentrated on the description of multimedia content. While previous MPEG standards have focused on the coded representation of audio-visual content, MPEG-7 is primarily concerned with secondary textual information about the audio-visual content [4, 5].

1.3 Video Coding Basics

1.3.1 Need for Video Coding

The main problem with the uncompressed (raw) video is that it contains immense amount of data and hence communication and storage capabilities are limited and expensive. Table 1.2 shows the raw data rates of a number of typical video formats, whereas Table 1.3 shows a number of typical video applications and the bandwidths available to them.

Consider a 2-h CCIR-601 color movie. Without compression, a 5-Gbit compact disk (CD) can hold only 30 s of this movie. To store the entire movie on the same CD requires a compression ratio of about 240:1. Without compression the same movie will take about 36 days to arrive at the other end of a 384 kbits/s Integrated Services Digital Network (ISDN) channel.

This proves that the amount of uncompressed video data is too large for limited transmission bandwidth or storage capacities. It is also evident that the biggest challenge is to reduce the size of the video data using video compression. For this

Table 1.2 Raw data rates of typical video formats

Format	Raw data rate
HDTV	1.09 Gbits/s
CCIR-601	165.89 Mbits/s
CIF @ 15 f.p.s.	18.24 Mbits/s
QCIF @ 10 f.p.s.	3.04 Mbits/s

Table 1.3 Typical video applications

Application	Bandwidth
HDTV (6 MHz channel)	20 Mbits/s
Desktop video (CD-ROM)	1.5 Mbits/s
Videoconferencing (ISDN)	384 kbits/s
Videophone (PSTN)	56 kbits/s
Videophone (GSM)	10 kbits/s

reason, the terms "video coding" and "video compression" are often used inter-changeably, which are essential components of digital video services.

1.3.2 Elements of a Video Coding System

An aim of video coding is to reduce or compress the number of bits used to represent video. Video signals contain three types of redundancy: statistical, psychovisual, and coding redundancy. Statistical redundancy is present because certain data patterns are more likely than others. This is mainly due to the high spatial (intraframe) and temporal (interframe) correlations between neighboring pixels. Psychovisual redundancy is due to the fact that the human visual system (HVS) is less sensitive to certain visual information than to other visual infor-mation. If video is coded in a way that uses more and/or longer code symbols than absolutely necessary, it is said to contain coding redundancy. Video compression is achieved by reducing or eliminating these redundancies.

Figure 1.1 shows the main elements of a video encoder block. Each element is designed to reduce one of the three basic redundancies. The mapper (or trans-former) transforms the input raw data into a representation that is designed to reduce statistical redundancy and make the data more amenable to compression in later stages. The transformation is a one-to-one mapping and is, therefore, reversible.

Quantizer reduces the accuracy of the mappers output, according to some fidelity criterion, in an attempt to reduce psychovisual redundancy. This is a many-to-one mapping and is, therefore, irreversible. It represents a set of continuous-valued samples with a finite number of symbols. If each input sample is quantized independently, the process is referred to as scalar quantization. If, however, input samples are grouped into a set of vectors and this set is mapped to a finite number of vectors, the process is known as vector quantization. Scalar quantization techniques are involved in most image and video coding standards with the combination of transform coding. Another key element of video coding systems is the symbol encoder. This assigns a codeword to each symbol at the output of the quantizer. The symbol encoder must be designed to reduce the coding redundancy present in the set of symbols. Commonly used symbol encoding techniques are run-length coding, arithmetic coding, and Huffman coding. An output of the quantization step may contain long runs of identical symbols. Run-length encoding (RLE) is used to reduce this redundancy. RLE can represent such runs with intermediate symbols of the form (RUN, LEVEL). In Huffman coding there is a one-to-one correspondence between the symbols and code words. In arithmetic

Fig. 1.1 Elements of a video encoder

coding, however, a single variable-length codeword is assigned to a variable-length block of symbols. Huffman coding and arithmetic coding is widely used for image and video data.

1.4 Intraframe Coding

Intraframe coding refers to video coding techniques that achieve compression by exploiting (reducing) the high spatial correlation between neighboring pixels within a video frame. Such techniques are also known as spatial redundancy reduction techniques or still-image coding techniques. Intra prediction and transform coding are some of the techniques used for intraframe coding.

Predictive coding was originally proposed by Cutler in 1952 [6]. In the intra prediction method, a number of previously coded pixels are used to form a prediction of the current pixels. The difference between the pixels and its prediction forms the signal to be coded. Obviously, the better the prediction, the smaller the error signal, and the more efficient the coding system. At the decoder, the same prediction is produced using previously decoded pixels, and the received error signal is added to reconstruct the current pixels. Predictive coding is commonly referred to as differential pulse code modulation (DPCM).

Transform coding, developed more than two decades ago, has proven to be a very effective video coding method to reduce the spatial redundancy. In the 'spatial domain' (i.e., the original form of the image), samples are highly spatially correlated. The aim of transform coding is to reduce this correlation, ideally leaving a small number of visually significant transform coefficients (important to the appearance of the original image) and a large number of insignificant coefficients (that may be discarded without significantly affecting the visual quality of the image).

When choosing a transform, three main properties are desired: good energy compaction, data-independent basis functions, and fast implementation. The Karhunen-Loeve transform (KLT) is the optimal transform in an energy compaction sense. Unfortunately, this optimality is due to the fact that the KLT basis functions are dependent on the covariance matrix of the input block. Recomputing and transmitting the basic functions for each block is a nontrivial computational task. These disadvantages severely limit the use of the KLT in practical coding systems. The performance of many suboptimal transforms with data-independent basis functions have been studied [7]. Examples are the discrete Fourier transform (DFT), discrete cosine transform (DCT), Walsh-Hadamard transform (WHT), and the Haar transform. It has been demonstrated that the DCT has the closest energy-compaction performance to that of the optimum KLT [7]. Due to attractive features like near-optimum energy-compaction, data-independent basis functions, and fast algorithms the DCT has become the "workhorse" of most image and video coding standards.

1.5 Interframe Coding

Video is a time sequence of still images or frames. Thus, a naive approach to video coding would be to employ any of the still images (or intraframe). The coding methods discussed in Sect. 1.4 are on a frame-by-frame basis. However, the compression that can be achieved by this approach is limited because it does not exploit the high temporal correlation between the frames of a video sequence. Interframe coding refers to video coding techniques that achieve compression by reducing this temporal redundancy. For this reason, such methods are also known as temporal redundancy reduction techniques. Note that interframe coding may not be appropriate for some applications. For example, it would be necessary to decode the complete interframe coded sequence before being able to randomly access individual frames. Thus, a combined approach is normally used in which a number of frames are intraframe coded (I-frames) at specific intervals within the sequence and the other frames are interframe coded (predicted or P-frames) with reference to those anchor frames. In fact, some systems switch between interframe and intraframe within the same frame.

1.5.1 Three-Dimensional Coding

The simplest way to extend intraframe image coding methods to interframe video coding is to consider 3-D waveform coding. For example, in 3-D transform coding based on the DCT, the video is first divided into blocks of M $\times$ N $\times$ K pels (M; N; K denote the horizontal, vertical, and temporal dimensions, respectively). A 3-D DCT is then applied to each block, followed by quantization and symbol encoding, as illustrated in Fig. 1.2. It requires K frame memories both at the encoder and decoder to buffer the frames. In addition to this storage requirement, the buffering process limits the use of this method in real-time applications because encoding/decoding cannot begin until all of the next K frames are available.

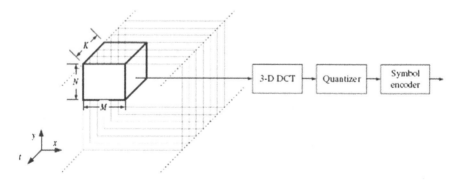

Fig. 1.2 A 3-D transform coding system

1.5.2 Interframe Predictive Coding

In this process compression is achieved with two main functions:

1. *Prediction.* Create a prediction of the current frame based on one or more previously transmitted frames.
2. *Compensation.* Subtract the prediction from the current frame to produce a 'residual frame'.

The output of this process is a residual (difference) frame and the more accurate the prediction process, the lesser the energy contained in the residual frame. The residual frame is encoded by the symbol encoder and sent to the decoder which recreates the predicted frame, adds the decoded residual, and reconstructs the current frame. This is *interframe coding* where frames are coded based on some relationship with other video frames, i.e., coding exploits the interdependencies of video frames. The entire process is shown in Fig. 1.3.

1.5.3 Frame Differencing

The simplest method of temporal prediction is to use the previous frame as the predictor for the current frame. Two successive frames from a video sequence are shown in Fig. 1.4a, b. Frame 1 is used as a predictor for Frame 2 and the residual formed by subtracting the predictor (Frame 1) from the current frame (Frame 2) is shown in Fig. 1.4c. In this image, mid-gray represents a difference of zero and light or dark gray correspond to positive and negative differences respectively. It is clear that much of the residual data is zero; hence, compression efficiency can be improved by compressing the residual frame rather than the current frame.

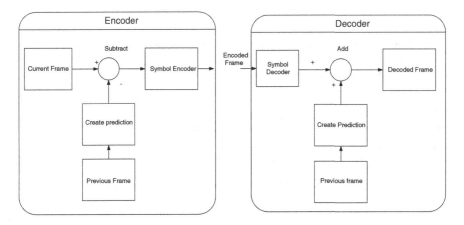

Fig. 1.3 Video codec with prediction

Fig. 1.4 a Current frame; **b** Previous Frame; **c** Residual frame (No motion compensation)

The decoder faces a potential problem that can be illustrated with Table 1.4 which shows the sequence of operations required to encode and decode a series of video frames using frame differencing. For the first frame the encoder and decoder use no prediction. The problem starts with Frame 2; the encoder uses the original Frame 1 as a prediction and encodes the resulting residual. However, the decoder only has the decoded Frame 1 available to form the prediction. Because the coding process is lossy, there is a difference between the decoded and original Frame 1 which leads to a small error in the prediction of Frame 2 at the decoder. This error will build-up with each successive Frame and the encoder and decoder predictors will rapidly 'drift' apart, leading to a significant drop in decoded quality.

The solution to this problem is for the encoder to use a decoded frame to form the prediction. Hence the encoder in the above example decodes (or reconstructs) Frame 1 to form a prediction for Frame 2. The encoder and decoder use the same prediction and drift should be reduced or removed. Figure 1.5 shows the complete encoder which now includes a decoding 'loop' in order to reconstruct its prediction reference. The reconstructed (or 'reference') frame is stored in the encoder and decoder to form the prediction for the next coded frame.

1.5.4 Motion Compensated Prediction

Frame differencing gives better compression performance than intraframe coding when successive frames are very similar, but does not perform well when there is a significant change between the previous and current frames. An improved

Table 1.4 Prediction drift

Encoder input	Encoder prediction	Encoder output/decoder input	Decoder prediction	Decoder output
Original frame1	Zero	Compressed frame1	Zero	Decoded frame1
Original frame2	Original frame1	Compressed residual frame2	Decoded frame1	Decoded frame2
Original frame3	Original frame2	Compressed residual frame2	Decoded frame2	Decoded frame3

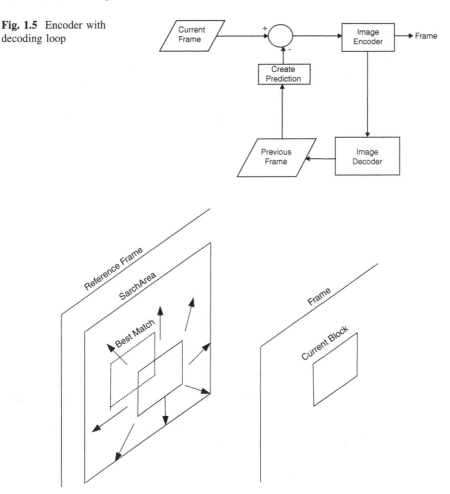

Fig. 1.5 Encoder with decoding loop

Fig. 1.6 Motion estimation

performance can be obtained by improving the prediction of changed regions. This can be achieved using motion estimation and compensation. Changes between frames are mainly due to the movement of objects. Using a model of the motion of objects between frames, the encoder estimates the motion that occurred between the reference frame and the current frame. This process is called motion estimation (ME) which is illustrated in Fig. 1.6.

An encoder then uses this motion model and information to move the contents of the reference frame to provide a better prediction of the current frame. This process is known as motion compensation (MC), and the prediction so produced is called the motion-compensated prediction (MCP) or the displaced-frame (DF). In this case, the coded prediction error signal is called the displaced-frame difference (DFD). A block diagram of a motion-compensated coding system is

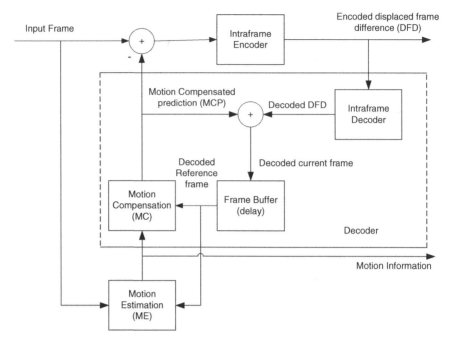

Fig. 1.7 Motion compensated video coding

illustrated in Fig. 1.7. This is the most commonly used interframe coding method. The reference frame employed for ME can occur temporally before or after the current frame. The two cases are known as forward prediction and backward prediction, respectively. The standards do not specify the encoder which performs the motion estimation. However, the standards give liberty to use different motion estimation techniques compatible to it. This allowed a vast scope for developing motion estimation techniques; this is why this field has attracted numerous researchers in the past two decades. The most commonly used ME method is the block-matching motion estimation (BMME) algorithm [8] which has emerged as the motion estimation technique achieving the best compromise between complexity and quality.

The research presented in this work is mainly concerned with improvements in classical motion estimation techniques for video coding.

1.6 Summary

Motion estimation is a key issue in the field of moving image analysis. While encoding the video frame, motion estimation and compensation are combined to exploit the spatio-temporal correlation of an image sequence along the motion trajectory.

This correlation is one of the most important compression factors of a video coder. The present research work mainly aims at improvements in classical motion estimation and compensation techniques, and thereby suggests novel ideas to improve the performance of video coding schemes.

References

1. CCIR. Recommendation 601-2: CCIR (currently ITU-R): Encoding parameters of digital television for studios, in *Digital Methods of Transmitting Television Information*, pp. 95–104, 1990
2. CCITT SG XV. Recommendation H.261–video codec for audiovisual services at p*64kbit/s. Technical Report COM XV –R37-E, August 1990
3. I. E. G. Richardson., *H.264 and MPEG-4 Video Compression* (Wiley, Chichester, 2003)
4. J. Hunter, An overview of the MPEG-7 description definition language (DDL. IEEE Trans. Circuits Syst. Video Technol. **11**(6), 765–772 (2001)
5. International Organization for Standardization. *ISO/IEC 15938-5:2003: Information Technology—Multimedia Content Description Interface—Part 5: Multimedia Description Schemes*, 1st edn. Geneva, Switzerland, 2003
6. C. C. Cutler, Differential quantization of communication signals U. S. Patent No. 2605361, 29 July 1952
7. R. J. Clarke, *Transform Coding of Images (Microelectronics and Signal Processing)* (Academic Press, London, 1985)
8. J.R. Jain, A.K. Jain, Displacement measurement and its application in interframe image coding. IEEE Trans. Commun. **29**(12), 1799–1808 (1981)

Chapter 2
Performance Evaluation of Block Matching Algorithms for Video Coding

A video sequence typically contains temporal redundancy; that is, two successive pictures are often very similar except for changes induced by object movement, illumination, camera movement, and so on. Motion estimation and compensation are used to reduce this type of redundancy in moving pictures. The block-matching algorithm (BMA) for motion estimation has proved to be very efficient in terms of quality and bit rate; therefore, it has been adopted by many standard video encoders. In this chapter, the basic principle of block matching motion estimation and compensation is introduced and fast motion search algorithms are addressed.

2.1 Search Algorithms for Motion Estimation

There exist two basic approaches to motion estimation:

1. Pixel-based motion estimation;
2. Block-based motion estimation.

The pixel-based motion estimation approach seeks to determine motion vectors for every pixel in the image. This is also referred to as the 'optical flow method', which works on the fundamental assumption of brightness constancy, that is, the intensity of a pixel remains constant when it is displaced. However, no unique match for a pixel in the reference frame is found in the direction normal to the intensity gradient. It is for this reason that an additional constraint is also introduced in terms of the smoothness of velocity (or displacement) vectors in the neighborhood. The smoothness constraint makes the algorithm interactive and requires excessively large computation time, making it unsuitable for practical and real-time implementation.

An alternative and faster approach is the block-based motion estimation. In this method, the candidate frame is divided into nonoverlapping blocks (of size

S. Metkar and S. Talbar, *Motion Estimation Techniques for Digital Video Coding*, SpringerBriefs in Computational Intelligence, DOI: 10.1007/978-81-322-1097-9_2, © The Author(s) 2013

16×16, or 8×8, or even 4×4 pixels in the recent standards) and for each such candidate block, the best motion vector is determined in the reference frame. Here, a single motion vector is computed for the entire block, whereby we make an inherent assumption that the entire block undergoes translational motion. This assumption is reasonably valid, except for the object boundaries. Block-based motion estimation is accepted in all the video coding standards proposed till date. It is easy to implement in hardware and real-time motion estimation and prediction is possible.

The effectiveness of compression techniques that use block-based motion compensation depends on the extent to which the following assumptions hold:

- The illumination is uniform along motion trajectories.
- The problems due to uncovered areas are neglected.

For the first assumption it neglects the problem of illumination change over time, which includes optical flow but does not correspond to any motion. The second assumption refers to the uncovered background problem. Basically, for the area of an uncovered background in the reference frame, no optical flow can be found in the reference frame. Although these assumptions do not always hold for all real-world video sequences, they continue to be used as the basis of many motion estimation techniques.

2.2 Principle of Block Matching Algorithm

The block matching technique is the most popular and practical motion estimation method in video coding. Figure 2.1 shows how the block matching motion estimation technique works. Each frame of size $M \times N$ is divided into square blocks $B\,(i, j)$ of size $(b \times b)$ with $i = 1....,\ M/b$ and $j = 1.....N/b$. For each block B_m in the current frame, a search is performed on the reference frame to find a matching based on a block distortion measure (BDM). The motion vector (MV) is the displacement from the current block to the best matched block in the reference frame. Usually, a search window is defined to confine the search. The same motion vector is assigned to all pixels within block.

$$\forall \vec{r} \in B(i,j), \quad \vec{d}(\vec{r}) = \vec{d}(i,j) \tag{2.1}$$

where the image intensity at pixel location $\vec{r} = (d_x, d_y)^T$ and at time t is denoted by $I(\vec{r}, t)$ and $\vec{d} = (d_x, d_y)^T$ is the displacement during the time interval Δt.

Suppose a block has size $b \times b$ pixels and the maximum allowable displacement of an MV is $\pm w$ pixels both in horizontal and vertical directions, there are $(2w + 1)^2$ possible candidate blocks inside the search window. The basic principle of block matching algorithm is shown in Figs. 2.2 and 2.3.

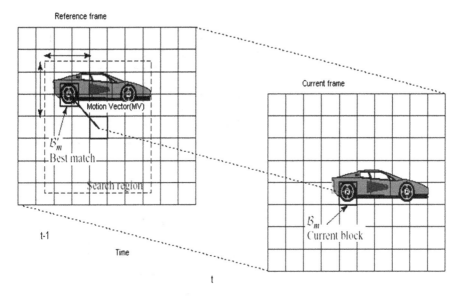

Fig. 2.1 Block matching motion estimation

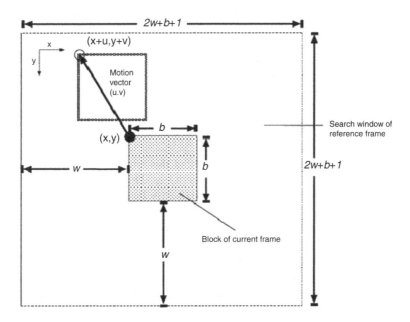

Fig. 2.2 Block matching method

A matching between the current block and one of the candidate blocks is referred to as a point being searched in the search window. If all the points in a search window are searched, the finding of a global minimum point is guaranteed.

Fig. 2.3 Search point in a
search window

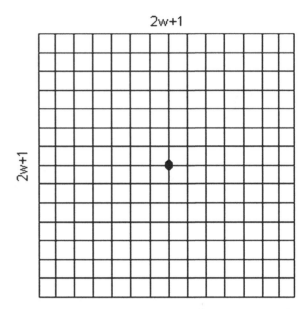

There are different parameters of the BMA with impact on performance and accuracy in motion estimation and compensation. The first important parameter is the distortion function, the other is the block size, and finally, the maximum allowed motion displacement, also known the search range. All these parameters are elaborated in the following sections.

2.2.1 Block Distortion Measure

In order that the compressed frame looks like the original, the substitute block must be as similar as possible to the one it replaces. Thus, a matching criterion or distortion function is used to quantify the similarity between the target block and candidate blocks.

Assume, F_t is the current frame and F_{t-1} is the reference frame. $F(x, y)$ is the intensity of a pixel at (x, y) in frame F. Each candidate block is located at $(x + w_x, y + w_y)$ inside a search window of size $\pm w$ pixels such that $-w \leq w_x, wy \leq +w$. The optimum motion vector which minimizes BDM function is (u, v). There are a number of criteria to evaluate the "goodness" of a match. Popular matching criteria used for block-based motion estimation are:

- Mean Square Error (MSE);
- Sum of absolute difference (SAD).

- **Mean Square Error (MSE)**

  The mean square error of a block of pixels computed at a displacement (w_x, w_y) in the reference frame is given by

$$MSE(w_x, w_y) = \frac{1}{N \times N} \sum_{i=x}^{x+N-1} \sum_{j=y}^{y+N-1} \left[F(i,j) - F_{t-1}(i + w_x, j + w_y) \right]^2 \qquad (2.2)$$

The MSE is computed for each displacement position (w_x, w_y) within a specified search range in the reference frame, and the displacement that gives the minimum value of MSE is the displacement vector which is more commonly known as motion vector and is given by

$$(u, v) = \min_{-w \le w_x, w_y \le +w} MSE(w_x, w_y) \qquad (2.3)$$

The MSE criterion defined in Eq. (2.2) requires computation of N^2 subtractions, N^2 multiplications (squaring), and $(N^2 - 1)$ additions for each candidate block at each search position. MSE is the Euclidian distance between current and reference blocks. It is considered to be better BDM because it is closer to our visual perception. The drawback of MSE is that it is more complex than other distortion measures as it needs square operations.

- **Sum of Absolute Difference (SAD)**

  Similar to the MSE criterion, the sum of absolute difference (SAD) too makes the error values as positive, but instead of summing up the squared differences, the absolute differences are summed up. The SAD measure at displacement (w_x, w_y) is defined as

$$SAD(w_x, w_y) = \sum_{i=x}^{x+N-1} \sum_{j=y}^{y+N-1} \left| F(i,j) - F_{t-1}(i + w_x, i + w_y) \right| \qquad (2.4)$$

The motion vector is determined in a manner similar to that for MSE as

$$(u, v) = \min_{-w \le w_x, w_y \le +w} SAD(w_x, w_y) \qquad (2.5)$$

The SAD criterion shown in Eq. (2.4) requires N^2 computations of subtractions with absolute values and N^2 additions for each candidate block at each search position. The absence of multiplications makes this criterion computationally more attractive for real-time implementation.

2.2.2 Block Size

Another important parameter of the block matching technique is the block size. If the block size is smaller, it achieves better prediction quality. This is due to a number of reasons. A smaller block size reduces the effect of the accuracy problem. In other words, with a smaller block size, there is less possibility that the block will contain different objects moving in different directions. In addition, a smaller block size provides a better piecewise translational approximation to non-translational motion. Since a smaller block size means that there are more blocks (and consequently more motion vectors) per frame, this improved prediction quality comes at the expense of a larger motion overhead. Most video coding standards use a block size of *16 × 16* as a compromise between prediction quality and motion overhead.

2.2.3 Search Range

The maximum allowed motion displacement '*w*' also known as the search range, has a direct impact on both the computational complexity and the prediction quality of the block matching technique. A small '*w*' results in poor compensation for fast-moving areas and consequently poor prediction quality. A large '±*w*' on the other hand, results in better prediction quality but leads to an increase in the computational complexity (since there are $(2w + 1)^2$ possible blocks to be matched in the search window). A larger '*w*' can also result in longer motion vectors and consequently a slight increase in motion overhead [1]. In general, a maximum allowed displacement of $w = \pm 7$ pixels is sufficient for low-bit-rate applications. MPEG standard uses a maximum displacement of about ±15 pixels, although this range can optionally be doubled with the unrestricted motion vector mode.

2.3 Full Search Algorithm

One of the first algorithms to be used for block-based motion estimation is full search algorithm (FSA), which examines exhaustively all positions in the search area. The FSA is optimal in the sense that if the search range is correctly defined, it is guaranteed to determine the best matching position. However, if the search range in either direction is '*w*' with step size of 1 pixel and assumes the search range is square, there are in total $(2w + 1)^2$ times of displacement in order to find a motion vector for each block, requiring a large amount of computations, especially for a large search window. The high computational requirements of FSA make it unacceptable for real-time software-implemented video applications. For real-time implementation, quick and efficient search strategies were explored.

2.4 Fast Block Matching Algorithms

Many fast search algorithms have been proposed to reduce the computational complexity of FSA while retaining similar prediction quality. All of them make use of the quadrant monotonic model [2]. The quadrant monotonic model, first used for block matching by Jain and Jain, assumes that the value of the distortion function increases as the distance from the point of minimum distortion increases. Therefore, not only the candidate blocks close to the optimal block better match than those far from it, but also the value of the distortion function is a function of the distance from the optimal position. Thus, the quadrant monotonic assumption is a special case of the principle of locality. The quadrant monotonic assumption allows for the development of suboptimal algorithms that examine only some of the candidate blocks in the search area. In addition, they use the values of the distortion function to guide the search toward a good match. As the entire candidate blocks are not examined, the match found might not be the best available. But the trade-off between the quality of the match and the number of matching criteria evaluations is usually good. Some of the popular fast block matching algorithms are discussed in the following sections.

2.4.1 Two-Dimensional Logarithmic Search Algorithm

The two-dimensional logarithmic search (TDL), introduced by Jain and Jain in 1981, was the first block-matching algorithm to exploit the quadrant monotonic model to match blocks [3]. The initial step size 's' is $\lceil w/4 \rceil$ (where $\lceil \rceil$ is the upper integer truncation function) where 'w' is the search range in either direction. The block at the center of the search area and the four candidate blocks at a distance 's' from the center on the x and y axes are compared to the target block to determine the best match. The five positions form a pattern similar to the five points of a Greek cross (+). Thus, if the center of the search area is at position [0, 0], then the candidate blocks at position [0, 0], [0, +s], [0, −s], [−s, 0], and [+s, 0] are examined. Figure 2.4 shows the search pattern of 2D-logarithmic search algorithm.

The step size is reduced by half only when the minimum distortion measure point of the previous step is the center (cx, cy) or the current minimum point reaches the search window boundary. Otherwise, the step size remains the same. When the step size is reduced to one, all eight blocks around the center position

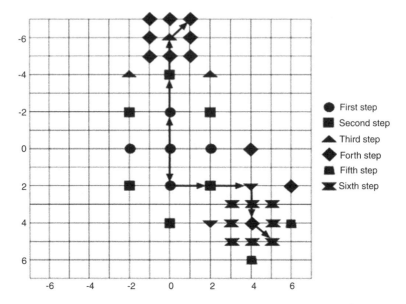

Fig. 2.4 2D-logarithmic search

which are $[cx - 1, cy - 1]$, $[cx - 1, cy]$, $[cx - 1, cy + 1]$, $[cx, cy - 1]$, $[cx, cy]$, $[cx, cy + 1]$, $[cx + 1, cy - 1]$, $[cx + 1, cy]$, and $[cx + 1, cy + 1]$ are examined, minimum distortion measure point of these is determined to be the best match for the target block and then it halts the algorithm. Otherwise (step size greater than one), the candidate blocks at positions $[cx, cy]$, $[cx + s, cy]$, $[cx - s, cy]$, $[cx, cy + s]$, and $[cx, cy - s]$ are evaluated for distortion measure. An experimental result proves that the algorithm performs well in large motion sequences because search points are quite evenly distributed over the search window.

2.4.2 Three-Step Search Algorithm

This algorithm is based on a coarse-to-fine approach with logarithmic decreasing in step size as shown in Fig. 2.5. The three-step search algorithm (TSS) tests eight points around the center [4].

For a center $[cx, cy]$ and step size 'd' the positions $[cx - d, cy - d]$, $[cx - d, cy]$, $[cx - d, cy + d]$, $[cx, cy - d]$, $[cx, cy]$, $[cx, cy + d]$, $[cx + d, cy - d]$, $[cx + d, cy]$, $[cx + d, cy + d]$ are examined. After each stage, the step size is halved and minimum distortion of that stage is chosen as the starting center of the next stage. The procedure continues till the step size becomes one. In this manner, TSS reduces the number of searching points as equal to $[1 + 8\{\log_2(d + 1)\}]$. One problem that occurs with the TSS is that it uses a uniformly allocated checking point pattern in the first step, which becomes inefficient for small motion estimation.

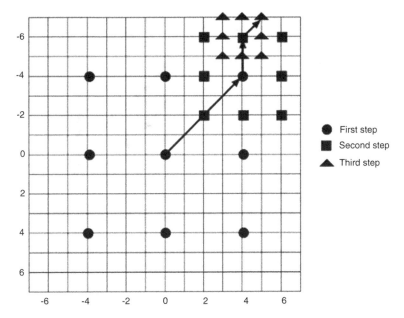

Fig. 2.5 Three-step search algorithm

2.4.3 Cross Search Algorithm

The cross search algorithm (CSA) proposed by Ghanbari [5] is a logarithmic step search algorithm using a saltire cross (×) searching patterns in each step. The CSA is presented in Fig. 2.6. The initial step size is half of maximum motion displacement 'w'. At each stage, the step size is halved, until the final stage is equal to one. At the final stage, however, the end points of a Greek cross (+) are used to search areas centered around the top-right and bottom-left corners of the previous stage, and a saltire cross (x) is used to search areas centered around the top-left and bottom-right corners of the previous stage.

The CSA requires $[5 + 4\lceil \log_2 w \rceil]$ comparisons where 'w' is the largest allowed displacement. The algorithm has a low computational complexity. It is, however, not the best in terms of motion compensation.

2.4.4 One-at-a-Time Search Algorithm

The One-at-a-time Search Algorithm (OTA) is a simple but effective algorithm, which has a horizontal and a vertical stage [6]. OTA starts its search at search window center. The center points and its two horizontally adjacent points, i.e., $(0, 0)$ $(0, -1)$ and $(0, +1)$ are searched. If the smallest distortion is for the center points, start the vertical stage, otherwise look at the next point in the horizontal direction

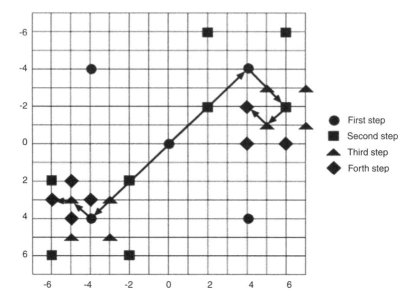

Fig. 2.6 Cross search algorithm

Fig. 2.7 One-at-a-time algorithm

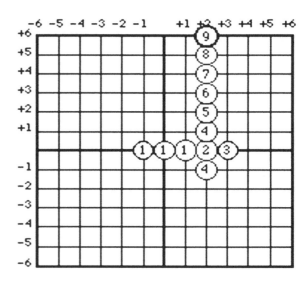

closer to the point with the smallest distortion, and continue in that direction till you find the point with the smallest distortion. The step size is always one. OTA stops when the minimum distortion point is closeted between two points with higher distortion. The above procedure is repeated in vertical direction about the point that has the smallest distortion in the horizontal direction. The search pattern of OTA is shown in Figure 2.7. This search algorithm requires less time, however, the quality of the match is not very good.

Performance evaluation of FSA, TDL, 3SS, CSA, and OTA algorithms in terms of quality and computational complexity is discussed in Sect. 2.7.

2.5 Proposed Modified Algorithms

2.5.1 New One-at-a-Time Algorithm

In this section, a modified version of OTA, called modified OTA, is proposed. It outperforms OTA in terms of computational complexity as compared to OTA algorithm. As compared to OTA, instead of evaluating in horizontal direction till the minimum distortion point is closeted between two points with higher distortion, the proposed algorithm checks four points around the center in the horizontal direction. Initial checking points are $(i, j - 1)$ $(i, j - 2)$ $(i, j + 1)$ and $(i, j + 2)$ around the center (i, j). The step size is always one. If optimum is found at the center, further procedure stops pointing motion vector as (i, j). This will save more than 80 % of the computational time. Otherwise, we proceed to search around the point in vertical direction where the minimum was found. Figure 2.8 illustrates the NOTA to find the positions of minimum distortion.

The steps of the proposed NOTA are explained as follows:

Step-I Evaluate the objective function for all five points in the horizontal direction.

Step-II If the minimum occurs at the center, stop the search; the motion vector points to the center.

Fig. 2.8 New one-at-a-time algorithm

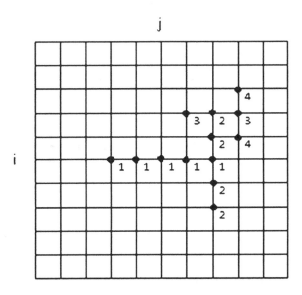

Step-III Otherwise, evaluate the objective function at four points on either side of previous (having minimum distortion of Step I winning point in the vertical direction around the winning point.).

Step-IV Search the two positions on either side of the winning point of step III horizontally; as regards the point that has the smallest distortion in the horizontal direction search the two positions vertically.

Step-V Minimum distortion point is declared as the best match.

The proposed NOTA retains the simplicity and regularity of OTA. Although NOTA uses more checking points in the first step as compared to OTA, preemption of the algorithm at the first step reduces the computations significantly. The experimental results of the proposed NOTA as compared to OTA are reported in Sect. 2.7.

2.5.2 Modified Three-Step Search Algorithm

The three-step search algorithm has been widely used in the block matching motion estimation due to its simplicity and effectiveness. However a TSS uses uniformly allocated checking point pattern in its first step, which is inefficient for the estimation of small motions, hence a modified three-step search algorithm is proposed. The features of MTSS are that it employs a center-biased checking point pattern in the first step, and a halfway-stop technique to reduce the computational cost.

The search procedure for MTSS is shown in Fig. 2.9. We have considered the search window size as ±7.

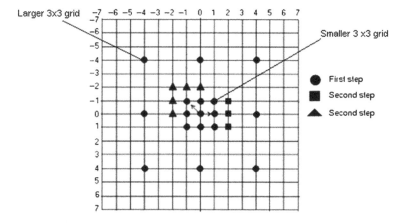

Fig. 2.9 Two different search paths of MTSS

The MTSS algorithm utilizes a smaller 3×3 grid search pattern at the center in addition to the larger 3×3 grid in the TSS. Thus distortion is evaluated for 17 search points. If the minimum BDM point is found at the center of the search window, the search will stop.

If the minimum BDM point is one of the eight points on the smaller 3×3 grid, only additional five or three points will be checked which depends on previous minimum distortion point. Otherwise, the search window center will be moved to the winning point on the larger 3×3 grid and the remaining procedure is same as in TSS. The detail of the algorithm is explained in the following steps.

Step-I For the first Step of MTSS along with larger 3×3 grid search points as in TSS, additional eight search points on smaller 3×3 grid at center, with step size equals to one are evaluated. This way, total $8 + 9 = 17$ search points needs to be evaluated in the first Step. If the minimum BDM point is the search window center, the search will be terminated; otherwise algorithm proceeds for Step II.

Step-II If one of the central eight neighboring points on the 3×3 grid is found to be the minimum in the first Step, go to Step III; otherwise go to Step IV.

Step-III Move the smaller 3×3 grid so that the window center is the winning point found in Step I. Evaluate additional five or three points according to the location of the previous winning point, and then the search is stopped declaring minimum BDM point as the winning point.

Step-IV Reduce the step size of larger 3×3 grid by half and move the center to the minimum BDM point in Step I, hence forward, procedure of TSS algorithm is followed till step size becomes one.

Figure 2.9 shows two different search paths for finding motion vector within 7×7 search area. According to the halfway-stop technique, the M3SS needs to evaluate 17 search points for stationary blocks and 20 or 22 points for small motion within central 3×3 search area. For the worst case $25 + 8 = 33$ points will be required, compared to 25 points in TSS. It has been experimentally proved that proposed MTSS shows good performance as compared to 3SS for slow motion sequence. The results of MTSS are discussed in Sect. 2.7.

2.6 Video Sequences for Simulation

In this research work, ten standard Quarter Common Intermediate File Format (QCIF) video sequences of different motion contents are used for performance comparison of different algorithms. These video sequences are categorized into three classes; Class A, Class B, and Class C, with increasing motion complexity.

| Silent | Claire | Grandma |

Fig. 2.10 First frames of video sequences "silent", "claire", and "grandma"

| News | Suzie | Miss America |

Fig. 2.11 First frames of video sequences "news", "suzie", and "miss" America

That is, the video sequences in Class A have low or slow motion activities. Those in Class B have medium motion activities and Class C videos have high or complex motion activities. The video sequences of Silent, Claire, and Grandma are all of slow object translation with low motion activities and belong to Class A. Their first frames are shown in Fig. 2.10. The first frames of the Class B sequences, News, Suzie, and Miss America with moderate motion are revealed in Fig. 2.11. Similarly, Fig. 2.12 presents the first frames of Foreman, Carphone, Salesman, and Trevor sequence having fast object translation with high motion activity which belongs to Class C.

2.7 Experimental Results

In this section, the performance of FSA, 3SS, 2-D logarithmic search, and OTA is discussed along with the proposed algorithm from the viewpoint of prediction of accuracy as well as the computational complexity.

Foreman   Carphone

Salesman   Trevor

Fig. 2.12 First frames of video sequences foreman, carphone, salesman, and trevor

2.7.1 Experimental Setup and Performance Evaluation Criterion

All the algorithms have been tested on desktop computer P-IV 2.4 GHz CPU. In our simulation block distortion measure (BDM) is defined to be the Mean Square Error (MSE). The block size is considered as 8×8 as tradeoffs between computational complexity and quality. The maximum motion in rows and column is assumed to be ± 7. Analysis has been done using three standard video sequences in QCIF (176×144) format, each representing different class of motion. These include **Silent, News,** and **Foreman.** The first 100 frames of the above-mentioned sequences have been used for simulation.

The quality of the reconstructed sequence should be estimated by subjective tests. One of the subjective metrics is Mean Square Error (MSE) which is evaluated between original frame and reconstructed frame. The lesser the value of MSE, the better the prediction quality. Mean Square Error is given by

$$MSE(i,j) = \frac{1}{M \times N} \sum_{m=1}^{M} \sum_{n=1}^{N} (f(m,n) - f'(m,n))^2 \qquad (2.6)$$

where $f(m, n)$ represents the current frame and $f'(m, n)$ is the reconstructed frame with frame size as M × N. Another widely used metric for comparing various image compression techniques is the peak-signal-to-noise-ratio (PSNR). The mathematical formulae for PSNR is

$$PSNR = 10\log_{10}(\frac{(2^b - 1)^2}{MSE})$$ (2.7)

The b in the equation is the number of bits in a pixel. For 8-bit uniformly quantized video sequence $b = 8$. The higher the value of PSNR, the better the quality of the compensated image.

Within the same search window, FSA can achieve the highest prediction quality because it searches all possible search points in the search window and thus is guaranteed to find optimum global minimum point. Therefore, prediction measure achieved by FSA is often used as reference to compare or evaluate other fast block matching algorithms. We consider the computational complexity of the algorithm in terms of the CPU time required to terminate the evaluation of the search algorithm. The time required increases linearly with the number of points in the search window being searched.

2.7.2 Performance Comparisons of Fast Block Matching Algorithms

In this section, we present experimental results to evaluate the performance of the FSA, TDLS, CSA, and OTA along with the proposed algorithms from the viewpoint of computational complexity as well as prediction accuracy.

From Table 2.1 it is found that for all sequences FSA shows higher PSNR values as compared to other algorithms with increased CPU time. MTSS has quality improvement in terms of PSNR over 3SS except for Foreman sequence. For example, the PSNR of MTSS is 37.99 and 37.34 db, higher than that of 3SS for sequences Silent and News, respectively. Coarse searches in TDLS and 3SS can locate the rough position of the global minimum and subsequent four searches can find the best motion vector. They perform well in large motion sequence because search points are evenly distributed over the search window. For higher motion, Foreman sequence the PSNR for TDLS is 33.44 and 33.27 db for 3SS which is higher than other algorithms apart from FSA. However, TDLS required higher CPU time for computation as compared to other fast algorithms. OTA performs one-dimensional gradient descending search on the error surface twice. Although it desires less computational time as compared with other fast block matching algorithms, its prediction quality is low which is reflected in the PSNR entries. This is because one-dimensional gradient descend search is insufficient to provide a correct estimation of the global minimum position. With slight improvement in the speed as compared to OTA, the proposed NOTA achieves

Table 2.1 Performance of block matching algorithms

Algorithms	Avg. MSE	Avg. PSNR	Avg. CPU Time in Sec.
Using silent sequence			
FSA	12.54	37.86	15.70
3SS	14.84	37.31	1.15
OTA	27.57	36.29	0.37
CSA	18.44	36.41	0.48
TDLS	15.97	37.11	1.43
NOTA	23.02	36.63	0.33
MTSS	14.42	37.34	0.76
Using news sequence			
FSA	17.10	38.14	15.70
3SS	18.86	37.88	1.26
OTA	24.89	37.53	0.27
CSA	22.88	37.59	0.44
TDLS	20.05	37.91	1.39
NOTA	22.95	37.47	0.22
MTSS	18.82	37.99	0.78
Using foreman sequence			
FSA	26.36	34.22	15.70
3SS	34.13	33.27	1.11
OTA	42.49	31.59	0.50
CSA	50.98	30.79	0.52
TDLS	32.06	33.44	1.58
NOTA	41.65	32.22	0.41
MTSS	38.14	33.20	0.92

higher PSNR quality for slow to fast motion sequence. The PSNR of OTA is 36.29 and 31.59 db for **Silent** and **Foreman** sequence and for NOTA, PSNR values are 36.63 and 32.22, respectively. In comparison with the other fast block matching algorithms, while the computational complexity of the CSA is the second lowest, its compensation performance is not.

Figure 2.13 shows comparisons of the MSE and PSNR per frame for the first 100 frames by using various search algorithms (including FSA, 3SS, TDLS, CSA, OTA, and proposed NOTA and MTSS) for **Foreman**, **News**, and **Silent** sequence.

2.8 Summary

In this chapter the importance of motion compensation in video coding is reviewed. To find motion vectors, block matching motion estimation (BMME) is performed. BMME takes up to 70 % for the total encoding time in modern video coding standards. The simplest algorithm for BMME is full search algorithm which can also achieve the best matching quality. However, the computational complexity

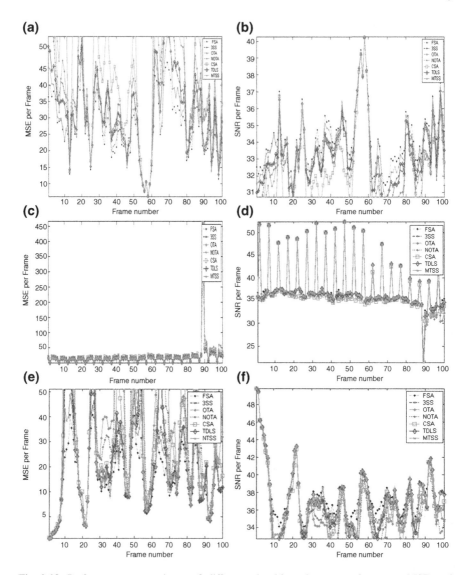

Fig. 2.13 Performance comparisons of different algorithms in terms of average MSE and average PSNR: (**a**) (**b**) Foreman (**c**) (**d**) News (**e**) (**f**) Silent

of full search algorithm is so high that it is unsuitable for many applications, e.g., real-time encoding. Fast block matching algorithms are proposed to reduce the computational complexity of FSA while maintaining similar matching quality. Some well-known algorithms which include 3SS, TDLS, CSA, and OTA are analyzed. To compare the performance of fast block matching algorithms, both matching quality and computational complexity have been considered. Matching quality is measured by Mean Square Error and Peak-Signal-to-Noise Ratio.

Computational complexity is measured by time required for motion estimation. Finally, video sequences used for analysis and simulation in this work are illustrated.

References

1. B. Liu, A. Zaccarin, New fast algorithms for the estimation of block motion vectors. IEEE Trans. Circuits Syst. Video Technol. **3**, 440–445 (1995)
2. International Organization for Standardization. *ISO/IEC 15938-5:2003: Information Technology—Multimedia Content Description Interface—Part 5: Multimedia Description Schemes*, 1st edn. Geneva, Switzerland, 2003
3. J.R. Jain, A.K. Jain, Displacement measurement and its application in interframe image coding. IEEE Trans. Commun. **29**(12), 1799–1808 (1981)
4. T. Koga, T. Ishiguro, Motion compensated inter-frame coding for video conferencing, *Proceedings of National Telecommunication Conference, New Orleans,* pp. G5.3.1–G5.3.5, Dec 1981
5. M. Ghanbari, The cross search algorithm for motion estimation. IEEE Trans. Commun. **38**(7), 950–953 (1990)
6. R. Srinivasan, K. R. Rao., Predictive coding based on efficient motion estimation. IEEE Trans. Commun. **33**(8), 888–896(1985)

Chapter 3
Fast Motion Estimation Using Modified Orthogonal Search Algorithm for Video Compression

Recently, a fast search algorithm for video coding using orthogonal logarithmic search algorithm (OSA) was proposed by Soongsathitanon et al. [1]. We have introduced the center-biased search point pattern for the estimation of small motions and a halfway-stop technique to reduce the computational complexity in the existing OSA. This chapter presents a novel method called modified orthogonal search algorithm (MOSA) for block-based motion estimation. Experimental results based on a number of video sequences are presented to demonstrate the advantages of the proposed motion estimation technique.

3.1 Introduction

The recent years have witnessed the vital role of video compression in data storage and transmission. Video compression involves the removal of spatial and temporal (interframe) redundancies. Interframe compression exploits the similarities between successive frames, known as temporal redundancy, to reduce the volume of data required to describe the sequence. Several interframe compression methods [2] of various degrees of complexity are presented in the literature such as subsampling coding, difference coding, block-based difference coding, and motion compensation. The motion compensation prediction is commonly used in most video codecs [2, 3]. In order to carry out motion compensation, the motion of moving objects has to be estimated first, which is called motion estimation [4]. Motion compensated prediction assumes that the current picture can be locally modeled as a translation of the pictures of some previous time. Block matching methods due to their less computational complexity are the most popular motion estimation methods, which are adopted by various video coding standards such as MPEG-1 and MPEG-2 [5]. In block matching method, the current frame is divided into sub-blocks of $N \times N$ pixels. Each sub-block is predicted from the previous or future frame, by estimating the amount of motion of the sub-block called motion vector during the frame time interval. The video coding syntax specifies how to

S. Metkar and S. Talbar, *Motion Estimation Techniques for Digital Video Coding*, SpringerBriefs in Computational Intelligence, DOI: 10.1007/978-81-322-1097-9_3, © The Author(s) 2013

represent the motion information for each sub-block, but it does not specify how such vectors are to be computed. The motion vector is obtained by finding a BDM function which calculates the mismatch between the reference and the current block [6]. To locate the best matched sub-block that produces the minimum mismatch error, we need to calculate distortion function at several locations in the search range.

One of the first algorithms developed for block-based motion estimation is the full search algorithm (FSA) or exhaustive search algorithm (ESA), which evaluates the block distortion measure (BDM) function at every possible pixel location in the search area [6]. Although this algorithm is the best in terms of quality of the predicted frame and simplicity, it is computationally intensive. In the past two decades, several fast search methods for motion estimation have been introduced to reduce the computational complexity of block matching, for example, two-dimensional logarithmic search (2DLS) [6], three-step search (3SS) [7], and four-step search (4SS) [8]. As these algorithms utilized uniformly allocated search points in their first step, they could achieve substantial computational reduction with the drawback of modest estimation accuracy degradation. Search with a large pattern in the first step is inefficient for the estimation of small motion since it will be trapped into a local minimum. In real-world video sequences, the distortion of motion vectors is highly center biased, which results in a center-biased motion vector distribution instead of a uniform distribution. This indicates that the probability increases to get the global minimum at the center region of the search window. To make use of this characteristic, center-biased block matching algorithms were then proposed with search points much closer to the center, which improved the average prediction accuracy, especially for slow motion sequences. Well-known examples of this category are new three-step search (N3SS) [9], advanced center-biased three-step search (ACBTSS) [10], block-based gradient descent search (BBGDS) [11], diamond search (DS) [12], cross diamond search (CDS) [13], hexagon-based search (HS) [14], etc. There is considerable research effort being applied to the subject of motion estimation [5, 15, 16, 17, 18].

Among these algorithms, 3SS has become the most popular for low bit rate application owing to its simplicity and effectiveness. However, 3SS and the recently proposed OSA use uniformly allocated searching points in their first step which becomes inefficient for the estimation of small motions since it gets trapped into local minimum. Having observed this problem, a modified orthogonal search algorithm (MOSA) is proposed here. The proposed MOSA searches the additional central eight points in order to favor the characteristics of center-biased motion. To speed-up block matching process, MOSA terminates at intermediate step instead of adapting the entire steps of the algorithm referred as halfway-stop technique.

The rest of the chapter is organized as follows: Sect. 3.2 presents the OSA, the proposed MOSA is described in Sects. 3.3, and 3.4 gives the experimental results of our proposed method in comparison with FSA, 3SS, and OSA. The experimental results are concluded in Sect. 3.5.

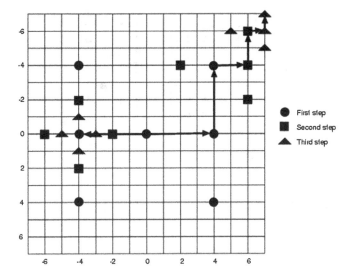

Fig. 3.1 Orthogonal logarithmic search

3.2 Orthogonal Logarithmic Search Algorithm

The proposed OSA [1] has pairs of horizontal and vertical search points with a logarithmic decrease in the successive step size (*'st'*). After step size *st* has been initialized at $st = \lceil d/2 \rceil$, where *'d'* is the maximum motion displacement, the center block and two candidate blocks on either side of the x-axis at a distance (*'±st'*) are compared to the target block. The position of minimum distortion (BDM) will become the center of the vertical stage. During the vertical stage, two search points above and below the new center are examined and the values of the distortion function at these three positions are compared. The position with the minimum distortion will become the center of the next stage. After horizontal and vertical iteration, the step size is halved, if it is greater than one the algorithm proceeds with another horizontal and vertical iteration, otherwise it halts and declares one of the positions from the vertical stage as the best match for the target block. Thus, OSA estimates motion vectors by searching the window orthogonally with logarithmic reduction in the step size. The search patterns for OSA are shown in Fig. 3.1.

3.3 Modified Orthogonal Search Algorithm

We present the modification in the original OSA by utilizing a 3 × 3 grid search pattern at the center to exploit the center-biased characteristics of motion vector distribution in real-world video sequences. Figure 4.2 shows the search pattern of MOSA. The search window size is (2d + 1) × (2d + 1), where *d* is the maximum

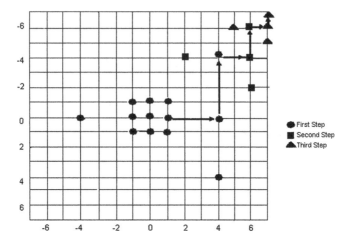

Fig. 3.2 Modified orthogonal logarithmic search

displacement in the horizontal and vertical directions. The step size is initialized at $st = \lceil d/2 \rceil$, where $\lceil \ \rceil$ is the upper truncation function. We have assumed the search window size as 7×7 (i.e., maximum displacement of motion vector d $= \pm 7$).

The proposed MOSA is decomposed into the following steps:

Step-I   In the initial stage of the algorithm, we propose to add eight extra search points around the center in addition to the original search points in horizontal direction used in OSA as shown in Fig. 3.2. The advantage of adding these extra search points is to capture the small motion at the center. If the minimum BDM at this stage occurs at the search window center, i.e., at (0, 0), then assume the block is with zero motion and further search comes to an end (halfway-stop technique). Otherwise go to step II.

Step-II   If the winning point of step I corresponds to one of the eight points of the center grid, then step size is initialized to one and two positions horizontally on either side and two vertical positions above and below of the winning point are searched. The search point with minimum BDM corresponds to motion vector. The algorithm is halted at this stage or else go to step III.

Step-III Shift the center at the winning point of step I (having minimum BDM) for the vertical search and two positions are searched in the vertical direction (up and down) retaining the same step size.

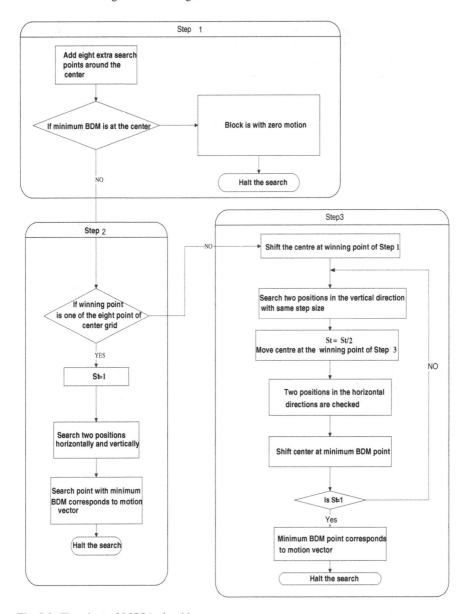

Fig. 3.3 Flowchart of MOSA algorithm

Step-IV   The step size '*st*' is reduced by half and the center is moved to the minimum BDM point of step III. Again two positions in the horizontal direction are checked with step size *st* apart from the center.

Step-V   Repeat steps III and IV, till step size '*st*' becomes one.

The proposed algorithm is graphically represented in Fig. 3.3. The proposed MOSA retains the simplicity and regularity of OSA. Although MOSA takes eight additional search points at the initial stage as compared to OSA, the early termination of the algorithm at step I helped to reduce the computations significantly.

3.3.1 Analysis of the Algorithm

For block matching algorithms, the number of search points required for each motion vector estimation measures the computational complexity. In this section, we estimate the number of search points needed to find motion vector for MOSA as compared with the OSA and 3SS algorithms. The number of search points required for 3SS algorithm given by [7] is,

$$[1 + 8\{\log_2 (d + 1)\}] \tag{3.1}$$

Hence, 3SS requires a total of 25 search points with 7×7 search window. The computational complexity of the OSA algorithm measured by [1] is,

$$[1 + 4 \lceil \log_2(d + 1) \rceil] \tag{3.2}$$

For 7×7 search window size, the OSA algorithm requires 13 search points. A search example using the center-biased MOSA for motion vector estimated at $(-7, 7)$ is shown in Fig. 3.2. The 21 search points are required to obtain the motion vector for this particular example. From Fig. 3.2 we can easily deduce that the total number of search points can vary from 11 (best-case scenario) to 21 (worst-case scenario). However, experimental results indicate that average search points for 100 frames are less than 11, since for the boundary blocks the search window extends outside the frame boundary, and as a result some search points get omitted from the search pattern. From this discussion, we can undoubtedly claim that, with MOSA, motion vectors are found with fewer search points as compared to 3SS and OSA algorithms.

3.4 Experimental Results

The algorithm is implemented on a Pentium IV, 2.4 GHz personal computer. In our experiment BDM is defined to be the mean square error (MSE) [19].

$$MSE(i,j) = \frac{1}{N^2} \sum_{m=1}^{N} \sum_{n=1}^{N} (f(m,n) - f(m+i, n+j))^2 - d \leq i, j \leq d \tag{3.3}$$

where $f(m, n)$ represents the current sub-block of N^2 pixels at coordinates (m, n) and $f(m + i, n + j)$ represents the corresponding block in the previous frame at new coordinates $(m + i, n + j)$. Choosing large block size is computationally

Table 3.1 Average MSE for different algorithms

Video sequences	Algorithms			
	FSA	3SS	OSA	MOSA
Foreman	26.36	34.13	40.2	31.23
Claire	3.60	3.67	3.96	3.61
Carphone	20.6	24.52	26.6	22.61
Silent	12.54	14.87	16.42	17.90
News	17.10	18.86	19.98	19.98
Grandma	3.70	3.78	3.92	3.72
Suzie	15.09	18.32	22.33	20.42
Miss America	5.02	5.54	5.79	5.10
Salesman	6.51	6.97	7.43	6.82
Trevor	30.08	35.47	38.27	38.33

Table 3.2 Average SNR for different algorithms

Video sequences	Algorithms			
	FSA	3SS	OSA	MOSA
Foreman	34.2	33.2	32.6	33.59
Claire	43.25	43.20	42.99	43.24
Carphone	35.2	34.6	34.32	34.95
Silent	37.86	37.18	36.81	36.76
News	38.14	37.95	37.75	37.88
Grandma	43.65	43.59	43.53	43.62
Suzie	36.71	36.01	35.33	35.88
Miss America	41.54	41.18	41.02	41.48
Salesman	40.86	40.64	40.45	40.76
Trevor	35.55	35.21	34.87	35.33

efficient but results in the poor quality of the reconstructed frame, hence we choose block size 8×8 as a tradeoff between computational complexity and the quality of the image [20]. The maximum displacement of block in horizontal and vertical directions is assumed to be $d = \pm 7$. For the performance evaluation of standard algorithms FSA, OSA, 3SS, and MOSA, 100 frames of ten QCIF video sequences Foreman, Claire, Carphone, Silent, News, Grandma, Suzie Salesman, Trevor, and Miss America are selected. These video sequences are categorized into three classes based on fast, moderate, and slow motion. The video sequences of Silent, Claire Grandma are all of slow object translation with low motion activities. News, Suzie, and Miss America are with moderate motion while Foreman, Carphone, Salesman, and Trevor sequences are of fast object translation with high motion activity. The motion compensated values of MSE and SNR are used as metrics to evaluate the performance of standard algorithms and the proposed MOSA with results are presented in Tables 3.1 and 3.2.

Computational complexity in terms of average number of search points per block is summarized in Table 3.3.

Table 3.3 Average search points for different algorithms

Video sequences	Algorithms			
	FSA	3SS	OSA	MOSA
Foreman	205.18	25	12.40	11.14
Claire	205.18	25	12.39	10.66
Carphone	205.18	25	12.40	11.43
Silent	205.18	25	12.39	10.62
News	205.18	25	12.39	10.41
Grandma	205.18	25	12.40	10.63
Suzie	205.18	25	12.40	11.26
Miss America	205.18	25	12.40	11.06
Salesman	205.18	25	12.39	10.37
Trevor	205.18	25	12.39	10.76

It is seen from Table 3.3 that the proposed MOSA algorithm requires less number of search points as a result of halfway-stop technique and it also improves the MSE and SNR values as compared to OSA which is shown in Tables 3.1 and 3.2. For Foreman and Carphone which contains large motion, the MOSA algorithm has obtained speed improvement of nearly 50 % over 3SS. For Claire and Grandma sequences with motion vectors limited within a small region, the proposed MOSA algorithms achieve close to 55 % speed improvement over 3SS as seen from Table 3.3. Tables 3.1 and 3.2 indicate that MOSA has given performance close to FSA for Claire and Grandma sequences with small motion content. Also, it performs better for Foreman, Salesman, and Carphone sequences of large motion over 3SS and OSA. Algorithms 3SS and OSA utilize uniformly allocated search points in their first stage, which is not efficient to catch small motions appearing at the search window center. By utilizing 3×3 patterns of search points, at the center of search window we are able to catch the small motion with MOSA. The advantage of the novel MOSA in terms of the speed of operation (computational complexity) is much more dramatic. Experimental results have shown that the MOSA performs better than OSA in terms of motion compensation error and it is superior to 3SS in terms of computational complexity. For more clarification, the performance measured in terms of MSE and SNR for three test sequences, each one representing a different class of motion like Foreman, Claire, and Miss America, are represented in Figs. 3.4, 3.5, and 3.6 respectively.

3.5 Summary

In this chapter, Modified Orthogonal Logarithmic Search (MOSA) is proposed in block motion estimation which results in significant speed-up gain over 3SS and FSA. Search for the center-biased motion vectors have been facilitated by adding

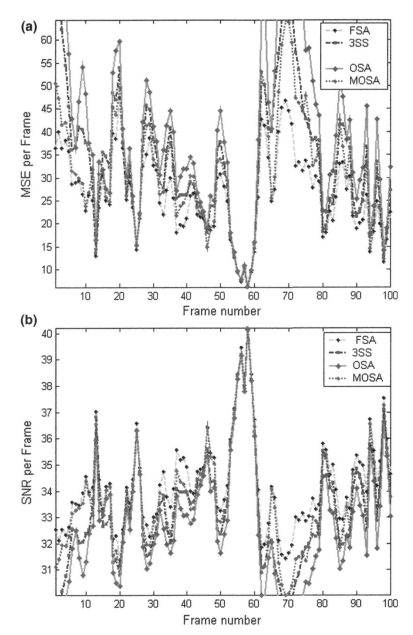

Fig. 3.4 Comparative MSE and SNR performance of "Foreman" using FSA, 3SS, OSA, and MOSA

eight extra search points, which are neighbors of the window center. The introduction of the halfway-stop technique reduced the computational complexity and the time required for the computation. The strength of the MOSA algorithm lies in

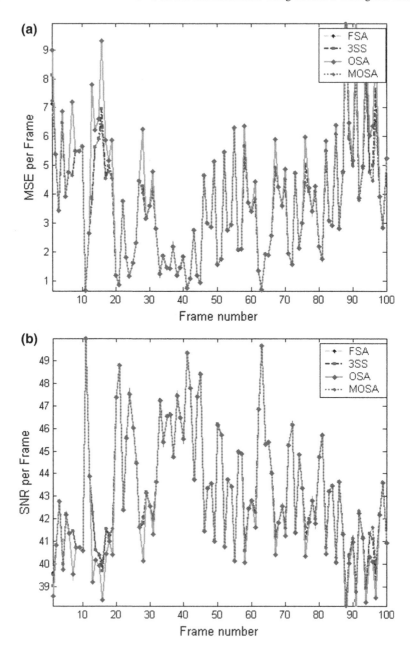

Fig. 3.5 Comparative MSE and SNR performance of "Claire" using FSA, 3SS, OSA, and MOSA

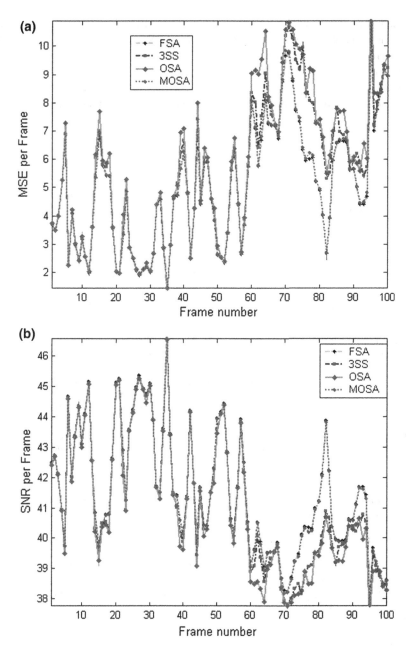

Fig. 3.6 Comparative MSE and SNR performance of "Miss America" using FSA, 3SS, OSA, and MOSA

its speed of operation which helps to reduce computational complexity. The speed of operation is almost 80 and 50 % faster than FSA and 3SS respectively. The SNR performance ensures the good quality of reconstructed frame comparable to those of FSA, 3SS, and OSA. Summing up, we conclude that MOSA is a good candidate to replace FSA and 3SS methods. Owing to its ability to reduce complexity, it can be recommended for efficient hardware implementation.

References

1. S. Soongsathitanon, W.L. Woo, S.S. Dlay, Fast search algorithms for video coding using orthogonal logarithmic search algorithm. IEEE Trans. Consum. Electron. **51**(2), 552–559 (2005)
2. ISO/IEC 11172-2, *Information Technology-Coding of Moving Picture and Associated Audio for Digital Storage Media at up to About* 1.5 mbit/s: *Part 2 Video*, Aug 1993
3. T. Koga, T. Ishiguro, Motion compensated inter-frame coding for video conferencing, *Proceedings of National Telecommunication Conference, New Orleans,* pp. G5.3.1–G5.3.5, Dec 1981
4. R. Srinivasan, K. R. Rao, Predictive coding based on efficient motion estimation. IEEE Trans. Commun. **33**(8), 888–896 (1985)
5. S. Zhu, K.K. Ma, A new diamond search algorithm for fast block matching motion estimation. IEEE Trans. Image Process. **9**(2), 287–290 (2000)
6. J.R. Jain, A.K. Jain, Displacement measurement and its application in interframe image coding. IEEE Trans. Commun. **29**(12), 1799–1808 (1981)
7. L. Yeong -Kang, A memory efficient motion estimator for three step search block-matching algorithm. IEEE Trans. Consum. Electron. **47**(3), 644–651 (2001)
8. L.M. Po, W.C. Ma, A novel four-step search algorithm for fast block motion estimation. Trans. Circuits Syst. Video Technol. **6**(3), 313–317 (1996)
9. H. Nisar, T. S. Choi, An advanced center biased three step search algorithm for motion estimation, *Proceedings of IEEE International Conference on Multimedia and Expo.* vol. 1, pp. 95–98, 2000
10. R. Li, B. Zeng, M.L. Liou, A new three-step search algorithm for block motion estimation. IEEE Trans.Circuits Syst. Video Technol. **4**(4), 438–442 (1994)
11. L.K. Liu, E. Feig, A block-based gradient descent search algorithm for block motion estimation in video coding. Trans Circuits Syst Video Technol. **6**(4), 419–422 (1996)
12. J.Y. Tham, S. Ranganath, M. Ranganath, A.A. Kassim, A novel unrestricted center-biased diamond search algorithm for block motion estimation. IEEE Trans. Circuits Syst. Video Technol. **8**(4), 369–377 (1998)
13. C.H. Cheung, L.M. Po, A novel cross-diamond search algorithm for fast block motion estimation. IEEE Trans. Circuits Syst. Video Technol. **12**(12), 16–22 (2002)
14. C. Zhu, X. Lin, L.P. Chau, Hexagon-based search pattern for fast block motion estimation. IEEE Trans. Circuits Syst. Video Technol. **12**, 349–355 (2002)
15. Lai-Man Po, Chi Wang Ting, Ka-HO Ng, New enhanced hexagon based search using point–oriented inner search for fast block motion estimation. Int. J. Signal Process. **3**(3), 193–197 (2006)
16. W.G. Hong, T.M. Oh, Sorting-based partial distortion search algorithm for motion estimation. Electron. Lett. **40**(2), 113–115 (2004)
17. L.-P. Chau, C. Zhu, A fast octagon –based search algorithm for motion estimation. J. Signal Process. **83**, 671–675 (2003)

18. H. Gharavi, M. Mills, Block matching motion estimation algorithms—new results. IEEE Trans. Circuits Syst. **37**(5), 649–651 (1990)
19. Y. Jar-Ferr, H. Shu-Sheng, L. Wei-Yuan, Simplified block matching criteria for motion estimation. IEICE Trans. Info. Syst. **E83-D**(4), 922–930 (2000)
20. A. Ahmad, N. Khan, S. Masud, M.A. Maud, Efficient block size selection in H.264 video coding standard. Electron. Lett. **40**, 19–21 (2004)

Chapter 4
Dynamic Motion Detection Techniques

A novel motion estimation technique with reduced complexity is introduced in this chapter. This is based on segmenting the current frame into active and inactive regions containing moving objects and stationary background, respectively. The inactive region is assumed with zero motion vectors. The early detection of blocks due to zero motion vector, leads to cut redundant computation significantly, which speeds up the coding of video sequences. New approach, based on frame variance and correlation, is proposed to decide the motion activity of a block. Block-based motion estimation has shortcomings like unreliable motion vectors, block artifacts, and poor motion-compensated prediction along the moving edges. A new search strategy based on edge detection is proposed to obtain more accurate motion prediction along moving edges.

4.1 Introduction

The wide range of multimedia applications based on video compression (video telephony, video surveillance, digital television) leads to different kind of requirements for a video coding standard (image quality, compression efficiency). Several multimedia application areas require high power efficiency (especially in the video encoder part) in order to work on embedded systems and mobile terminals [1, 2]. This requirement implies the need to dramatically reduce the complexity of the video encoder. Algorithmic analysis shows that motion estimation is the most complex module in the video encoder. This is mainly due to the great number of calculation in motion estimation. Having this in mind, the present chapter postulates dynamic motion detection for fast and efficient video coding [3]. Many algorithms are suggested in the literature to reduce the number of calculation intended for motion estimation based on spatial and temporal correlation of motion vector [4, 5, 6].

S. Metkar and S. Talbar, *Motion Estimation Techniques for Digital Video Coding*, SpringerBriefs in Computational Intelligence, DOI: 10.1007/978-81-322-1097-9_4, © The Author(s) 2013

We propose motion detection technique which is based on segmenting the current frame into blocks containing moving objects (an active region) and stationary background (an inactive region). The blocks detected by proposed technique as corresponding to stationary region are assumed with zero motion vectors. The early detection of blocks with zero motion vector leads skip significant redundant computation of motion estimation and thus speed up the coding of video sequence. A more precise threshold value is proposed to decide motion activity of a block.

Section 4.2 describes the proposed approach of motion detection, followed by motion estimation technique in Sect. 4.3. While Sect. 4.4 brings about experimental results and concluding touch is presented in Sect. 4.5.

4.2 Dynamic Motion Detection

There are two ways to reduce the computational cost of motion estimation. The first one is to speed up the algorithms themselves for which numerous fast algorithms have been proposed. The other way is to terminate the motion estimation calculation preemptively. Generally, for most sequences, there exist a significant number of blocks which results with a zero MV after motion estimation. If we can predict the zero-motion blocks (ZMB), we can stop motion estimation early and eliminate a portion of the computation. In the proposed technique motion detection is carried out prior to motion estimation to avoid the heavy computational overhead. Here we suggest two approaches based on frame variance and frame correlation which is considered as a threshold to decide the motion activity of frame blocks.

To estimate the blocks as zero motion blocks using dynamic motion detection technique an image is first divided into blocks of $n \times n$ pixels, where 'n' is usually set to 8. The blocks resulting from the partition of the current frame and the previous frame are called the current and the previous blocks, respectively. For each current block 'X', the technique will decide the motion activity of the block corresponding to the block in the previous frame with the same spatial position as 'X' based on the threshold decision.

4.2.1 Variance

Motion detection is achieved by thresholding the variance of each block. The variance provides more useful information for motion detection. Our method is based on the fact that the variance of total image and the image blocks cannot be closely matched unless their local means and variances are closely matched. This implies that blocks whose local means and variances differ greatly cannot constitute a close match. By sorting the blocks according to their variances and defining an acceptance criterion with respect to threshold as image variance to decide the motion activity, we can limit the number of blocks a relatively small for motion estimation.

We first define a temporal-activity measure based on the variance of the difference of two successive frames. The main idea behind the proposed technique is that it locates the rough position of a moving object based on the temporal-activity measure.

$$v_{FD} = \frac{1}{M \times N} \sum_{i=1}^{M} \sum_{j=1}^{N} \left(f_{FD}(i,j) - \bar{f}_{FD} \right)^2 \qquad (4.1)$$

where v_{FD} is the variance of the frame difference (FD). The local (block) variance of the block difference is given by

$$v_{BD} = \frac{1}{n \times n} \sum_{i=1}^{n} \sum_{j=1}^{n} \left(f_{BD}(i,j) - \bar{f}_{BD} \right)^2 \qquad (4.2)$$

where $\bar{f}_{BD}$ denotes the mean of the block difference of current and reference frame. The above equation indicates that the larger the variance of a block, as compared to the entire variance v_{FD}, the greater its temporal-activity. For each block, we compare its motion activity v_{BD} with the threshold v_{FD}. If $v_{BD} > v_{FD}$, the block is marked as a part of moving objects and if $v_{BD} < v_{FD}$, the block is marked as a part of stationary region. For analysis of the proposed technique, we selected three QCIF (176×144) video sequences News, Foreman, and Silent from a suite of standard test sequences having a broad range of different video input characteristics which can be elaborated as follows.

News (Fig. 4.1a) is sequence located inside a news broadcast room. A male and female newscaster sitting in the foreground read from their manuscripts into the camera, while in the background, between them, a rectangular screen displays video clips illustrating their news report.

The video on the background screen is less sharp and flashes between a scene of two dancers on stage and a close-up scene of the one twirling female ballerina. The white letters "MPEG4 WORLD" appear in both the lower left corner of the frame and one of the background TV monitors. *News* was chosen because of its contrasts. It combines the fast, rotational motion of the dancers with the minimal motion of the newscasters; it contrasts the sharply defined newsroom against the dancing scenes. News is also the only sequence with scene changes, which appear on the background screen.

Foreman (Fig. 4.1b) is a sequence that focuses on the talking head of a construction foreman, which moves up and down with the movement of the camera. The camera then zooms out to pan across a construction site. *Foreman* is considered to be a difficult sequence to encode efficiently.

Silent (Fig. 4.1c) sequence was chosen for its slow motion, significant amounts of panning. The sequence has moderate hand and head movements with stationary background.

Fig. 4.1 **a** News 4.1.
b Foreman 4.1. **c** Silent

Each video frame is partitioned into blocks of 8×8 pixels. In total, we have 396 blocks to be considered for the motion detection technique. As a representative, we have considered the two consecutive frames of the above mentioned sequence for the analysis. The variance of the current frame for the different blocks and the threshold, to decide the motion activity within frame obtained by the proposed technique is represented in Fig. 4.2a, b, c. It has been observed that with respect to threshold, most of the blocks for News and Silent sequence are

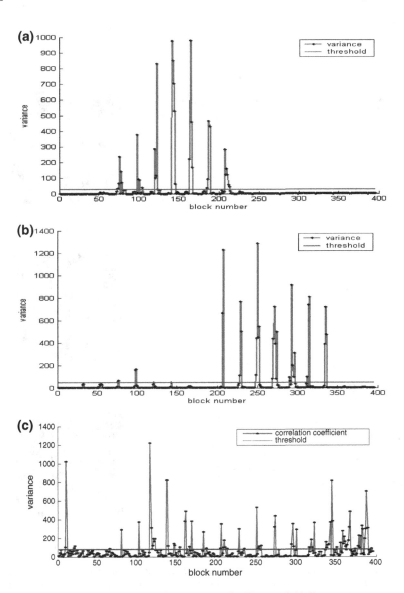

Fig. 4.2 Variance for different blocks for (**a**) News (**b**) Silent and (**c**) Foreman sequence

categorized as an inactive blocks. For **Foreman** the decision of blocks is little bit cumbersome.

The performance of the motion detection technique on the video sequences is represented by the Fig. 4.3. Figure 4.3a, b shows the two successive frames of **News, Silent,** and **Foreman** sequence. Figure 4.3c represent the motion detected region obtained by the proposed technique. The area without motion is indicated with black region.

The sequence **News** is with its background and object having significantly different characteristics. It can be seen that the technique extracted moving background efficiently except the lip motion of the male. The hand motion of

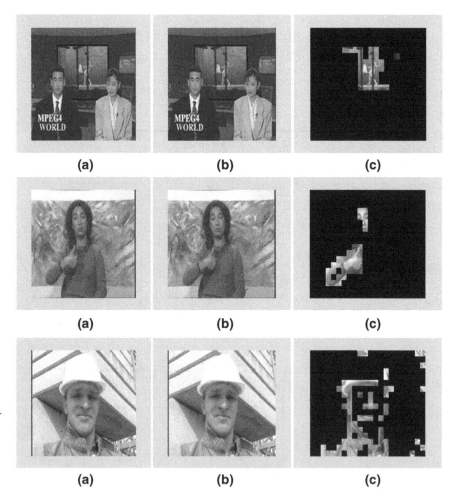

(a)	**(b)**	**(c)**
(a)	**(b)**	**(c)**
(a)	**(b)**	**(c)**

Fig. 4.3 a Original reference frame. **b** Original current frame of news, silent and foreman sequence. **c** Result of the variance based motion detection method: The blocks without motion are indicated with *Black*

Silent sequence is well identified by the proposed technique. **Foreman**, which is a little more complicated than **News** and **Silent**, but we preserve to determine the head motion of the sequence. It proves that the above procedure detects the moving blocks with significant temporal-activity.

The experiment indicates that the proposed technique identifies most of the moving blocks and also proficient to mark the zero motion blocks.

Next section describes the second approach based on the correlation.

4.2.2 Correlation

The correlation between two signals (cross correlation) is a standard approach for feature detection. Correlation function is widely used among known measures of similarity. Template matching or correlation is a technique that searches for specific features or characteristics within an image. This work describes the use of a feature correlation method for the reliable motion detection. A generic approach has been adopted to produce a threshold from images with different properties or characteristics. We use technique based on the correlation that uses level of similarity for comparison of corresponding fragments of the video sequence adjoining frames. The threshold value is set by the equation

$$cor = \frac{\sum_{i=1}^{M}\sum_{j=1}^{N}[A(i,j) - \overline{A}][B(i,j) - \overline{B}]}{\sqrt{\sum_{i=1}^{M}\sum_{j=1}^{N}[A(i,j) - \overline{A}]^2 \sum_{i=1}^{M}\sum_{j=1}^{N}[B(i,j) - \overline{B}]^2}} \tag{4.3}$$

where 'A' denotes the reference frame and B denotes the current frame. $M \times N$ is the frame size and $\overline{A}$, $\overline{B}$ are the mean of the reference and current frame, respectively. A method on moving block detection based on local correlation matching is developed, for which current and reference frame is partitioned into blocks of size $n \times n$. The local correlation match function is specified by

$$corb = \frac{\sum_{i=1}^{n}\sum_{j=1}^{n}[A_b(i,j) - \overline{A_b}][B_b(i,j) - \overline{B_b}]}{\sqrt{\sum_{i=1}^{n}\sum_{j=1}^{n}[A_b(i,j) - \overline{A_b}]^2 \sum_{i=1}^{n}\sum_{j=1}^{n}[B_b(i,j) - \overline{B_b}]^2}} \tag{4.4}$$

where A_b and B_b represents the mean of the block difference of current and reference frame respectively.

If the local correlation match function does not exceed the preset threshold 'cor' (i.e. $corb < cor$), the decision on presence of motion is made. And if the $corb > cor$ the block is categorized as a part of stationary background and is considered with zero motion vector. By using a correlation match function for motion detection it is shown that the method is capable to carry out stable motion detection.

The analysis of the proposed technique using correlation as threshold is carried on the similar sequence (**News, Foreman,** and **Silent**) which was described in the previous section. The graph of local correlation match function for the blocks and the threshold cost is plotted in Fig. 4.4. It has been observed that the local correlation match function for most of the blocks is less than the threshold which indicates that the blocks are not allied with correlation match function and hence these are categorized as motion (active) blocks. For **News** and **Silent** sequence around 75 % blocks are categorized as an inactive blocks. For **Foreman** sequence most of the blocks are categorized as an active blocks. In all, the result indicates that the proposed correlation technique accurately detects the motion within the sequence.

To investigate the motion detection performance, video sequence with different motion characteristics is considered. The motion detection results are shown in Fig. 4.5, which demonstrates the extracted object motion and stationary background which is indicated by black.

For the **News** sequence along with the background motion eye and lip motion of the male object is detected by the technique. For the same sequence, slight portion of the female object is falsely detected as motion area. The method performs well for **Silent** sequence to detect hand and head motion. For the **Foreman** sequence which includes the camera motion, most of the background region is correctly detected as an active region. The results indicate that motion detection performance is acceptable for all the types of video sequence to separate out an active and an inactive region.

4.3 Motion Estimation

The proposed motion detection module is used to separate out an active and an inactive region within current frame. Motion detection decides whether to subject the block of the current frame for the motion vector estimation or that block is judged with zero motion. The complexity of the video coder is strongly influenced by the number of calculations required to find motion vector. Consequently, by reducing the number of blocks for motion estimation process we can save a measurable time in the encoder process with minor effects on the quality of the produced video sequence. In this proposed work, three step search algorithm (3SS) owing to its simplicity and effectiveness is recommended for motion estimation of active region.

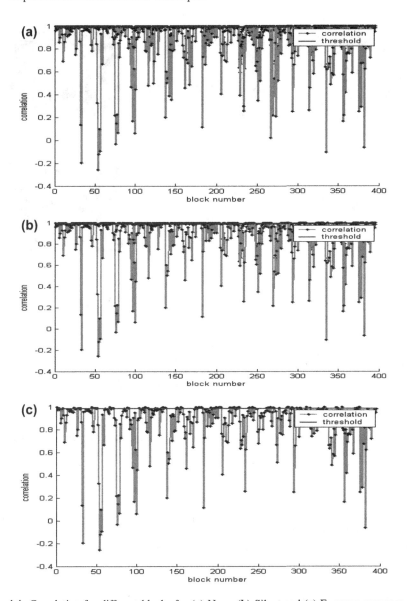

Fig. 4.4 Correlation for different blocks for (**a**) News (**b**) Silent and (**c**) Foreman sequence

4.4   Experimental Results of DMD Techniques

The test sequences used in the experimental results are the five standard QCIF
(176 × 144) videos (100 frames) of **Foreman**, **News**, **Grandma**, and **Claire**
defining different motion content. We had selected block size as 8 × 8 as a tradeoff
between computational complexity and the quality of the image. With this, each

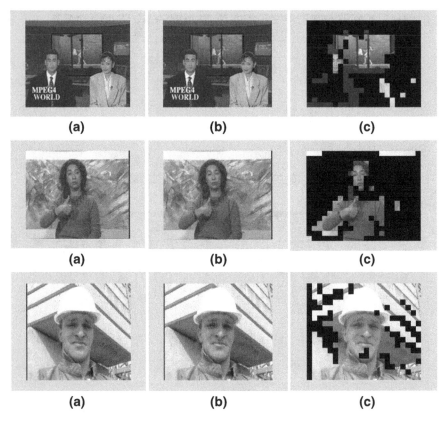

(a)	(b)	(c)
(a)	(b)	(c)
(a)	(b)	(c)

Fig. 4.5 **a** Original reference frame. **b** Original current frame of news, silent and foreman sequence. **c** Result of the correlation based motion detection method: The blocks without motion are indicated with black

frame is partitioned into total 396 blocks. With this method, some of the motion searches can be stopped early and the computation associated with these searches can therefore be reduced. Table 4.1 list the average number of an inactive blocks detected by the dynamic motion detection module. In these methods, the

Table 4.1 Average number of inactive blocks for different video sequences detected by MD technique

Threshold: Variance				
Foreman	News	Carphone	Grandma	Claire
290.350	348.930	306.9400	311.63	348.93
Threshold: Correlation				
Foreman	News	Carphone	Grandma	Claire
63.9400	260.7200	53.0500	253.1300	99.460

Table 4.2 Average MSE for different algorithms and different video sequences

Block matching algorithms	Video sequences				
	Foreman	News	Carphone	Grandma	Claire
FSA	26.36	17.10	20.6	3.70	3.60
3SS	34.13	18.86	24.52	3.78	3.67
MD + 3SS (Variance)	42.45	20.58	22.74	3.80	3.21
MD + 3SS (Correlation)	30.90	19.09	21.33	4.34	3.68

Table 4.3 Average PSNR for different algorithms and different video sequences

Block matching algorithms	Video sequences				
	Foreman	News	Carphone	Grandma	Claire
FSA	34.2	38.14	35.29	43.65	43.25
3SS	33.22	37.95	34.06	43.59	43.20
MD + 3SS (Variance)	32.83	37.12	35.03	43.15	43.33
MD + 3SS (Correlation)	33.47	37.80	35.12	42.11	43.24

thresholds are decided dynamically based on the frame content. Inactive blocks are considered with zero motion. As a result, numerous unnecessary calculations required for motion estimation of these blocks are saved. Experimental results show that our proposed method achieves significant complexity reduction, while the degradation in video quality is negligible.

Remarkable part of the algorithm is to indicate the valid search areas for motion estimation. With the knowledge of areas with motion, the motion estimation algorithm can be made to ignore blocks without motion. For every sequence, evaluation is done for Full Search Algorithm (FSA), Three-Step Search Algorithm (3SS) and finally the dynamic motion detection performance presented here along with 3SS algorithms. For the convenience of comparison, we have shown the average signal to noise ratio, average mean square error, and average search points needed for motion vector calculation.

Table 4.2 indicates that the performance of the proposed method degrades for the sequence of high motion content like **Foreman**. Where as the proposed technique shows good performance for the sequence of moderate to slow motion sequences like **Carphone** and **Claire**. Average peak signal to noise ratio is used as a measure of qualitative analysis which is shown in Table 4.3. From the results, we can see that the proposed methods are stable and work well for different sequences. It is seen that quality of the presented technique is close to the standard methods as FSA and 3SS. Graphical representation of PSNR performance is shown in Fig. 4.6.

An average number of search points measures the computational complexity since the computational time varies linearly with number of search points. The average number of search points for the different techniques is shown in Table 4.4. As compared with the fast motion estimation algorithm, the incorporation of DMD

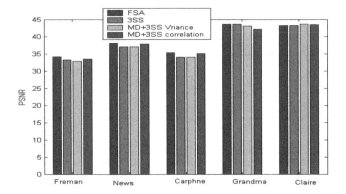

Fig. 4.6 Average PSNR for different algorithms

Table 4.4 Average search points for different algorithms

Algorithms	Video sequences				
	Foreman	News	Carphone	Grandma	Claire
FSA	205.8	205.8	205.8	205.8	205.8
3SS	25	25	25	25	25
MD + 3SS (Variance)	23.65	3.05	24.00	22.87	12.77
MD + 3SS (Correlation)	23	16.63	24.44	23.77	17.77

method reduces the search points. It has been observed that for fast motion sequence like Foreman, variance-based motion detection techniques shows nearly about 5.4 % speed improvement over 3SS and correlation-based technique shows 8 % speed improvement. Where as for slow motion sequence like Claire, speed improvement with respect to 3SS is approximately 50 and 28 % for variance and correlation technique, respectively.

Analysis shows that our proposed scheme can simplify the encoder complexity, maintaining good compression rate and fine video quality. The present algorithm is strongly influenced by the detection of threshold value, it is because the estimation of detected movements depends on threshold value (using high value, all of the motions cannot be detected; otherwise low value can cause the detection of background noise as relevant motion). In the projected method, threshold value is calculated dynamically and it is not taken as a constant value.

All the standard motion estimation algorithms can avail the advantage to reduce the complexity from the knowledge of areas with motion so all the types of motion estimation algorithms can be modified to benefit from the proposed motion detection techniques.

4.5 Summary

This chapter propounds the dynamic motion detection algorithm which is presented and tested on a set of sequences. DMD is an innovative algorithm for motion estimation complexity reduction based on active motion detection. A classifier based on the frame variance and correlation has been employed to detect active and inactive blocks. Using threshold as Variance is a better option for slow motion sequences and Correlation for fast to moderate motion sequences. Additionally, the threshold value in the algorithm is calculated automatically which avoids user interaction. It has been experimentally proved that the majorities of the blocks within a frame are classified as an inactive blocks and can be taken with zero motion vectors. As a result, we saved measurable computational time required for motion vector estimation for these blocks.

References

1. M. Paul, M. Murshed, L. Dooley, A real time pattern selection algorithm for very low bit-rate video Coding using relevance and similarity metrics. IEEE Trans. Circuits Syst. Video Technol. **15**(6), 753–761 (2005)
2. I.E.G. Richardson, *Video Codec Design* (Wiley, New York, 2002)
3. G. Bailo, M. Bariani, H.264 search window size algorithm for fast and efficient video coding with single pixel precision and no background estimation for motion detection, in *Proceedings of IEEE CCNC2006*, pp. 754–757, 2006
4. Y. Nakaya, H. Harashima, Motion compensation based on spatial transformations. IEEE Trans Circuits Syst. Video Technol. **4**(3), 339–356 (1994)
5. R. Korada, S. Krishna, Spatio-temporal correlation based ast motion estimation algorithm for MPEG-2, in *Proceedings of 35th IEEE Asilomar Conference on Signals, Systems and computers*, California, pp. 220–224, Nov 2001
6. J.Y. Nam, J.-S. Seo, J.-S. Kwak, M.-H. Lee, H.H. Yeong, New fast-search algorithm for block matching motion estimation using temporal and spatial correlation of motion vector. IEEE Trans. Consum. Electron. **46**(11), 934–942 (2000)

Chapter 5
Conclusion and Research Direction

This chapter concludes the book by summarizing the main developments and results of this work, and by indicating directions for further research.

5.1 Conclusion

As the Internet is becoming more and more universal and multimedia technology has progressed a lot, the communication of image data has become a part of life. In order to employ effect in a limited transmission bandwidth, to convey the most high quality user information, it is necessary to have more advanced compression method in image and data. Motion estimation (ME) and Compensation techniques, which can eliminate temporal redundancy between adjacent frames effectively, have been widely applied to popular video compression coding standards such as MPEG-2, MPEG-4. Motion estimation generates motion vectors that represent the movements of image blocks between successive frames. Motion estimation is conceptually quite simple and allows a very large reduction in the video bit rate. However, it often requires a considerably large part of the computation in video encoders. This research has addressed the developments in motion estimation techniques for image sequence coding applications. Block matching motion estimation techniques have been chosen as the baseline of this study due to their suitability for coding applications.

The book focused on the computational complexity and performance trade-offs offered by some well-known fast motion estimation algorithms in a video coding scheme such as Full Search Algorithm, Three-Step Search Algorithm, Two-Dimensional Logarithmic Search Algorithm, Cross-Search Algorithm, and One-at-a-Time Algorithm. New One-at-a-Time Algorithm and Modified Three-Step Search Algorithm are proposed with modifications. The New One-at-a-Time Search (NOTA) is a simple but effective algorithm. Experimental results show that the proposed technique provides competitive performance with reduced

S. Metkar and S. Talbar, *Motion Estimation Techniques for Digital Video Coding*, SpringerBriefs in Computational Intelligence, DOI: 10.1007/978-81-322-1097-9_5, © The Author(s) 2013

computational complexity. An efficient Modified Three-Step Search Algorithm was presented. The search strategy of M3SS performs better than 3SS and FS algorithm. Qualitative performance of M3SS in terms of average SNR does not show significant variation from FSA and 3SS algorithms. This analysis is useful for selecting appropriate algorithms for a variety of video coding applications that require an optimal trade-off between computational complexity and quality of video.

It has been observed that 3SS and recently proposed OSA use uniformly allocated searching points in their first step which becomes inefficient for the estimation of small motions since it gets trapped into local minimum. Having observed this problem, a novel modified orthogonal search algorithm (MOSA) is proposed in this book. The proposed MOSA searches the additional central eight points in order to favor the characteristics of center biased motion. To speedup block matching process, MOSA terminated at intermediate step instead of adapting the entire steps of the algorithm referred as halfway-stop technique. This feature improves the speed performance of the algorithm by 80 % as compared to the Full Search Algorithm, 50 % over the Three-Step Search algorithm, and 2 % faster than the OSA. The proposed modified orthogonal logarithmic search (MOSA) results in significant speed gain over 3SS and FSA. The strength of MOSA algorithm lies in its speed of operation. Owing to its ability of reducing complexity, we recommend MOSA for efficient hardware implementation.

The wide range of multimedia applications based on video compression (video telephony, video surveillance, digital television) leads to different kinds of requirements for a video coding standard (image quality, compression efficiency). Several multimedia application areas require high power efficiency (especially in the video encoder part) in order to work on embedded systems and mobile terminals. This requirement implies the need to dramatically reduce the complexity of the video encoder. Algorithmic analysis shows that motion estimation is the most complex module in the video encoder. This is mainly due to the great number of calculations in motion estimation. Having this in mind, we proposed dynamic motion detection techniques for fast and efficient video coding. Dynamic motion detection is an innovative algorithm for motion estimation and complexity reduction, based on active motion detection. A classifier based on the frame variance and correlation has been employed to detect active and inactive blocks. Additionally, the threshold value in the algorithm is calculated automatically which avoids user interaction. It has been experimentally proved that the majorities of the blocks within a frame are inactive and can be taken with zero motion vectors. As a result, we saved measurable computational time required for motion vector estimation for these blocks. The result shows that this proposed scheme can simplify the encoder complexity, maintaining high compression rate and good video quality.

5.2 Directions for Future Research

In the future, other evolutionary computing techniques can also be tried for better results. Three important factors like block size, search area, and distortion metric can be thought for improvement in the performance of Video Coder, such as variable block size, large search area for complex motions, and small search area for less complex motions.

The distortion between the original and compressed video sequences is typically evaluated as a mathematical error measurement, e.g., MSE, even though this type of measurement does not correlate well with perceptual quality and cannot lead to subjective optimal decisions. Therefore, objective distortion metrics considering perceptual quality, especially content-adaptive perceptual quality, are meaningful. The distortion metrics will play an important role in future video compression algorithm. Different acceptance rejection criteria can be set for different blocks during the matching process.

Multi-resolution technique is a good candidate for intraframe coding which is based on human visual perception. This may be computationally complex in today's scenario but looking into the growing research in computer architecture, this can be utilized in Video Coder.

New algorithms in motion estimation, segmentation, and tracking could be thought to allow coders to selectively transmit the only "interesting" parts of the video signal. In this respect, a lot of research can be carried out in order to stabilize the existing techniques and to further investigate the human perception of images. One could then conceive algorithms that are able to automatically characterize the content of images. The challenge of the new ISO MPEG-7 work item is of great interest here.

It is to be seen to what extent adaptability, context modeling, and other strategies can be developed into mature algorithms to compress video more efficiently than today's standardized coders. Future video compression algorithms may employ more adaptability, more refined temporal and spatial prediction models with better distortion metrics.

The cost to users is significant for implementation complexity at both the encoder and decoder. Fortunately, it seems bitrates have a slower doubling time than computing power, so the disadvantage of increasing implementation complexity may one day be balanced with much improved processor capabilities. Due to the advancement in the integration technology, designing of specialized processors for video compression can be thought of.

Although the imminent death of research into video compression has often been proclaimed, the growth in capacity of telecommunications networks is being outpaced by the rapidly increasing demand for services. The result is an ongoing need for better multimedia compression, particularly video and image compression. At the same time, there is a need for these services to be carried on networks of greatly varying capacities and qualities of service, and to be decoded by devices ranging from small, low-power, handheld terminals to much more capable fixed systems.

Hence, the ideal video compression algorithm should have high compression efficiency, be scalable to accommodate variations in network performance including capacity and quality of service, and be scalable to accommodate variations in decoder capability.